The Eye

The Eye:

Window to the World

By Lael Wertenbaker
and the Editors of U.S.News Books

U.S.NEWS BOOKS   Washington, D.C.

U.S.NEWS BOOKS

THE HUMAN BODY
The Eye:
Window to the World

Editor/Publisher: Roy B. Pinchot

Series Editor: Judith Gersten

Contributing Editor: Linda S. Glisson

Picture Editor: Leah Bendavid-Val

Book Design: David M. Seager

Art Consultant: Jack Lanza

Art Coordinators
Irwin Glusker, Kristen Reilly

Staff Writers
Christopher West Davis,
Kathy E. Goldberg, Karen Jensen,
Charles R. Miller, Doug M. Podolsky,
Matthew J. Schudel, Robert D. Selim,
Edward O. Welles, Jr.

Director of Text Research: William Rust

Chief Researcher: Bruce A. Lewenstein

Text Researchers
Allison Abood, Susana Barañano,
Barbara L. Buchman, Heléne Goldberg,
Michael C. McCarthy, E. Cameron
Ritchie, Loraine S. Suskind

Picture Researchers
Leora Kahn, David Ross,
Jean Shapiro Cantu, Greg Johnson,
Johanna Boublik

Art Staff
Esperance Shatarah, Raymond J. Ferry,
Martha Anne Scheele

Director of Production: Harold F. Chevalier

Production Coordinator: Diane B. Freed

Production Assistant: Mary Ann Haas

Production Staff
Carol Bashara, Ina Bloomberg,
Barbara M. Clark, Glenna Mickelson,
Sharon Turner

Quality Control Director: Joseph Postilion

Director of Sales: James Brady

Business Planning: Robert Licht

Fulfillment Director: Debra Hasday Fanshel

Fulfillment Assistant: Diane Childress

Cover Design: Moonink Communications

Cover Art: Paul Giovanopoulos

Author
Lael Wertenbaker is the author of 16 books
including the best seller *Death of a Man*, and
To Mend the Heart, the story of cardiac surgery.
Mrs. Wertenbaker, a foreign correspondent
and journalist during World War II, recently
joined a group of eminent cardiologists on a
two-week tour of the People's Republic of
China. Her reportage of that trip throws new
light on how Chinese medicine provides
health care for nearly a billion people.

Series Consultants
Donald M. Engelman is Molecular Biophysi-
cist and Biochemist at Yale University and a
guest Biophysicist at the Brookhaven National
Laboratory in New York. A specialist in bio-
logical structure, Dr. Engelman has published
research in American and European journals.
From 1976 to 1980, he was chairman of the
Molecular Biology Study Section at the
National Institutes of Health.

Stanley Joel Reiser is Associate Professor of
Medical History at Harvard Medical School
and codirector of the Kennedy Interfaculty
Program in Medical Ethics at the University.
He is the author of *Medicine and the Reign of
Technology* and coeditor of *Ethics in Medicine:
Historical Perspectives and Contemporary Concerns.*

Harold C. Slavkin, Professor of Biochemistry
at the University of Southern California,
directs the Graduate Program in Craniofacial
Biology and also serves as Chief of the
Laboratory for Developmental Biology in the
University's Gerontology Center. His research
on the genetic basis of congenital defects of
the head and neck has been widely published.

Lewis Thomas is Chancellor of the Memorial
Sloan-Kettering Cancer Center in New York
City. A member of the National Academy of
Sciences, Dr. Thomas has served on advisory
councils of the National Institutes of Health.
He has written *The Medusa and the Snail* and
The Lives of a Cell, which received the 1974
National Book Award in Arts and Letters.

Consultants for **The Eye**
Marc H. Bornstein is an Associate Professor
of Psychology at New York University. He
was awarded the American Psychological
Association's B. R. McCandless Young
Scientist Award in 1978. While his specific
interest is psychology, Dr. Bornstein has also
published material in the natural sciences,
the humanities and social sciences.

David Miller is Chief of Ophthalmology at
Beth Israel Hospital in Boston and Associate
Professor of Ophthalmology at Harvard
Medical School. Dr. Miller served as volunteer
ophthalmologist on Project Hope, the hospital
ship, during its tour of Colombia, South
America in 1967, and Tunisia in 1970. He has
written numerous books on ophthalmology
and received the Theo E. Obrig Memorial
Award for contact lens research in 1968.

Christian Wertenbaker is an ophthalmology
resident at the Albert Einstein College of
Medicine in New York. He has also served as
an instructor in neurology and ophthalmology
at that institution. In 1977, Dr. Wertenbaker
was a neuro-ophthalmology fellow at
Columbia-Presbyterian Medical Center in
New York City. He reviews articles for
Ophthalmology, the journal of the American
Academy of Ophthalmology.

Picture Consultants
Amram Cohen is General Surgery Resident at
the Walter Reed Army Medical Center in
Washington, D.C.

Richard G. Kessel, Professor of Zoology at
the University of Iowa, studies cells, tissues
and organs with scanning and transmission
electron microscopy instruments. He is
coauthor of two books on electron microscopy.

**U.S.News Books, a division of
U.S.News & World Report, Inc.**

**Library of Congress
Cataloging in Publication Data**

The Eye, window to the world.

(The Human body)
Includes index.
1. Eye. I. Wertenbaker, Lael Tucker, 1909–
II. Series.
QP475.5.E93   612'.84   81–16462
                                AACR2
ISBN 0–89193–603–3
ISBN 0–89193–633–5 (leatherbound)
ISBN 0–89193–663–7 (school ed.)

20 19 18 17 16 15 14 13 12 11
10   9   8   7   6   5   4   3   2   1

Contents

Introduction:
Light of the Mind 7

1 **So Great a Wonder** 9

2 **The Mechanics of Vision** 27

3 **Pictures in the Mind** 51

4 **Sensing Color** 77

5 **The Imperfect Eye** 101

6 **Illusion and Artifice** 125

Appendix 148
Glossary 152
Photographic Credits 155
Index 156

Introduction:

Light of the Mind

Allowing us to see into the harmony of life, light becomes electric when it passes through the eye. Looking from inside the eye, a camera gazes toward the sky through the expandable aperture of the pupil. Light refracted by the lens onto the retina is converted to electrical impulses that the brain interprets.

Sight is man's richest sense, his link to the world and its wealth of imagery. Vision begins with light, the abundant rain of the sun's energy falling through space to touch and warm the earth. Light projects value, tone and shadow into nature. The eye, keen to this kaleidoscopic effect, then relates what it senses to the brain, perception's ultimate seat.

Each waking second the eyes send some one billion pieces of fresh information to the brain. These fragments of light converge in the mind as images of stunning subtlety. The eyes can sense about ten million gradations of light and seven million different shades of color. Man sees in such detail because in the eye the diverse physical world comes closest to touching the raw perceptive power of the mind. The eye is an extension of the brain. It grows from that fertile organ like a plant reaching toward the light. In the back of the eyeball lies a patch of tissue no thicker than a postage stamp, no larger than a quarter. This is the retina, the light-catching net of the eye. It is woven of brain cells.

The retina can form, dissolve and create a new image every tenth of a second, an ability which matches man's power to lift the eye from the printed page, fix on a bird skimming across the horizon and then return to rejoice in the pattern of the wind through a nearby grove of trees. Quick to capture life's light-struck essence, the eye is also no less than a highly polished mirror of self. While emotions rise readily from the eye's lucid depths, so do signs of many bodily diseases. They darken the eye like storm clouds reflected in a still summer pond. Scientists peering into the eye find the curious riddle of a transparent organ shrouded in ample mystery. Though much has lately been learned of this complex wonder, it has scarcely begun to yield its manifold secrets. The eye promises to long hold man's inquiring gaze.

Chapter 1

So Great a Wonder

"Who would believe that so small a space could contain the images of all the universe?" So wrote the usually dispassionate observer Leonardo da Vinci. Studying the human eye roused him to awe: "O mighty process . . . what talent can avail to penetrate a nature such as this? What tongue will it be that can unfold so great a wonder? Truly none!"

Had Leonardo lived five centuries later, he would have known the many secrets of the eye that have yielded to science. In a sense, however, he was right. No single tongue could have unfolded the mysteries of the eye. It has taken the combined voices of philosophers, anatomists, mathematicians. For all his creative prowess, even the great Leonardo could not have dreamed the tale these voices have told.

The human eye has proved a worthy challenge to those who sought its secrets. Few creatures could boast of eyesight as powerful as man's. An astronaut orbiting the Earth could spot the pyramids of Egypt. Earthbound, he could also find minute flaws in a diamond.

But the eyes do more than inform the mind. They also betray the secrets of the heart. In the words of poet William Blake, they are "dim windows of the soul." Whether starry or dim, the eyes tell much about the human spirit that lies behind them. Puritan minister Cotton Mather sought to formalize the link between eye and soul into "moral diseases of the eye." An "unchaste eye," he asserted, became inflamed. Squinting revealed "an Aim at low and base Ends." Mather advised his readers to "employ your eye on the Book which will feed it well."

Warrior tribesmen in New Zealand, on slaying a chief, ate his eyes, for divinity rested there. Many cultures have credited the eyes with other supernatural powers. The evil eye, a malevolent influence, abounds in folklore. Even the New Testament lists it as a sin that pollutes man.

Hypnotic eyes of Hindu god Shiva stare from an eighteenth-century painting from India. Two youthful devotees brush flies from the trident, sacred weapon of the god. According to Hindus, Shiva is absolute god of the universe, capable of assuming many forms and reconciling life's paradoxes.

Watchful eyes of Egyptian deity Horus gaze from a limestone stele. God of the sun, Horus protected man from the evil god, Seth. In battling Seth, he lost an eye, which became a symbol of divine protection.

The power of the evil eye, capable of wreaking disease and death, pervaded Elizabethan superstition and found its way into Shakespeare's *Love's Labor's Lost*:

> Write, "Lord have mercy on us," on those three;
> They are infected, in their hearts it lies;
> They have the plague, and caught it of your eyes.

So persistent was the superstition that in 1881 the Russian Academy of Sciences decided to test it. The scientists starved a condemned prisoner for three days. Throughout the ordeal, he could see, but not touch, a loaf of bread. Afterward, they analyzed the bread and found it contained a "poisonous substance." History, however, does not record the precise nature of this toxin.

The notion endured alongside the awareness of the painful certainty that disease struck the eye itself. Among the first references to eye disease was the Code of Hammurabi, edicts of the famed ruler of the Babylonian Empire who reigned half a century, more than four thousand years ago. The prologue declared him a great soldier and pious king who sought to establish justice. In a Babylonian version of malpractice law, the Code called for a stiff penalty to those who damaged an eye in the course of operating on it: "He who has opened the eye of a free man with a brass needle and has destroyed it shall lose his hands." Historians believe the statute refers to primitive operations for abscesses, and possibly cataracts.

Medicine reached an unprecedented stage of refinement in Egypt, where its practitioners were men of wealth and high social standing. Their skill was renowned in the ancient world. Under the command of a high priest, magicians attempted to cure ailments through incantations. Another group, the sorcerers, used amulets and spells. A third class consisted of physicians, whose approach was more secular.

They were also the first specialists. "Medicine is practiced among them on a plan of separation," noted the well-traveled Herodotus in the fifth century B.C. "Each physician treats a single disorder, and no more," marveled the historian, "thus the country swarms with medical practitioners,

10

some undertaking to cure diseases of the eyes, others of the head, others again of the teeth, others of the intestines." Herodotus recorded still another variety of healer who treated diseases "which are not local." These healers, it seems, were also the first general practitioners.

The most remarkable record of Egyptian medicine was the Papyrus Ebers, a document describing all known diseases and remedies. Written around 1500 B.C., the papyrus was found in the mid-nineteenth century between the legs of a mummy near Thebes. Eight columns of the scroll were devoted to remedies for eye diseases. To "cure" blindness, the unknown author suggested the physician first "extract the water from the eyes of a hog," and mix it with "true collyrium, red lead and raw honey." This accomplished, the physician was instructed to "inject" the concoction in the ear of the patient.

Although the compiler of these unsavory recipes cited his sources, including "the priestly pharmacist Xui" and "a Jew from Byblos," he wisely left unrecorded the success rate of the cures. Primitive as they appear today, equally odious brews have been used throughout history. A few hundred years ago, the Pharmacopoeia of the Royal College of Physicians in London recommended preparations that included swallow's nest, powdered Egyptian mummy, pig bile and rhinoceros horn.

Invisible Atoms

Both Egyptians and Babylonians apparently limited themselves to treating eye diseases. They did not speculate on the nature of vision. Ancient Greeks, however, with their tremendous intellectual appetite, felt no such limitation. Democritus, a wealthy Thracian of the fifth century B.C., was among the first Greeks to study how the eye worked. He also explored music, ethics, mathematics and cosmology. Democritus proposed that the universe consisted of an infinite number of indivisible particles called atoms. Physical phenomena were produced by combinations of these invisible atoms that swarmed in empty space. "In our belief, there are sweet and bitter, hot and cold, in our belief there is color; but in truth," he declared, "there are atoms and void."

Hundred-eyed monster of Greek mythology, Argus was guardian to Hera, queen of the gods. Hera dispatched Argus to stand watch over the nymph, Io, a rival for Zeus's attentions. After Hermes beheaded Argus at Zeus's bidding, Hera adorned the peacock's tail with her slain sentinel's eyes.

On the bedrock of atomic theory, Democritus explained vision by arguing that objects shed their atoms into space. When atoms struck the eye, they were intercepted by the soul and thus perceived. Four centuries passed before Roman philosopher and poet Lucretius refined the atomic theory. "Amongst visible things many throw off bodies . . . as wood throws off smoke and fire heat," he reasoned, or "sometimes more closeknit and condensed, as . . . when the slippery serpent casts off his vesture among the thorns."

Rival Rays

Of all ancient Greek theories of vision, the atomic concept most closely conforms to modern ideas. But it never gained a wide audience in the Greek world. A rival theory deemed that the eye cast rays toward objects and so created sight. The theory was first stated formally by Empedocles, contemporary of Democritus. Greatly admired by Aristotle, Empedocles proposed that matter was composed of earth, air, fire and water. Comparing the eye to a lantern, he suggested that visual fire radiated from the pupil in the same way that light passed through a lantern's panes. According to legend, Empedocles thought himself a god. To prove his divinity, he hurled himself into the volcanic crater of Mt. Etna. The visual ray theory, however, survived to win him immortality of a different sort.

A generation later, Plato became an influential convert to the visual ray theory. Plato's prestige helped sustain the theory in the Middle Ages. But he deeply distrusted the physical senses, which he subordinated to the realm of reason. One story, probably apocryphal, tells of Plato's collecting the manuscripts of Democritus — a prime advocate of physical realities — only to burn them.

Despite his bias for abstraction, Plato grudgingly acknowledged a debt to the eye. He noted that the sight of "day and night, of months and the revolving years, of equinox and solstice" led to the invention of numbers, the notion of time, and the study of the world, "whence we have derived all philosophy." Contriving his own version of the visual ray theory, Plato suggested that the eyes' rays coalesced with daylight, and then

intercepted rays from objects. The interaction of rays and daylight produced vision. Color vision, he asserted, must be the result of encounters between particles in the visual rays and particles from objects. When the converging particles were the same size, the result was transparency; particles of different size produced colors. Whether or not Plato harbored misgivings about his theory, he cautioned against testing them by experiment.

Plato's pupil Aristotle chose not to heed the advice. Aristotle ventured into optics. Evolving his own ideas about vision, he firmly rejected the visual ray theory, reasoning that if the theory were true we would see in the dark. As further proof, he declared, "It is unreasonable . . . that the ray of vision reaches as far as the stars." Likewise, he set no store in Plato's coalescence theory, calling it "foolish."

Nevertheless, Aristotle accepted the premise that some kind of physical interaction occurred between the viewer and object. Otherwise, he noted, "There would be no need that either should occupy some particular place." Although complex, Aristotle's theory boiled down to the notion that the space between viewer and object conveyed the visual image.

Aristotle parted company with Empedocles and Plato when he dismissed the visual ray theory. But visual rays haunted optical theory for centuries. This was partly due to the enormous stature of the great mathematician Euclid, who wholeheartedly embraced the idea.

In *Optica*, Euclid applied geometry to the ray theory. Rays issued from the eyes in cones, he maintained, with the eye as vertex and object as base. Distant objects could not be seen because they fell between individual rays.

Borrowing from Greek theories, Romans made relatively few advances in optical theory — with the exception of second-century physician Galen, who thought the optic nerves were conduits of visual spirits, or "pneuma," from the brain. These spirits agitated the air near the eye. By contacting objects, the air produced sight. In short, the excited air became an extension of the eyes. Essentially an amplified version of earlier Greek thought, Galen's pneuma theory was overshadowed by the theories of Plato and Aristotle.

His contribution to optics lay in anatomy, and few improvements were made on his findings until Leonardo's time.

The Eye of Islam

Western science slumbered for centuries after the fall of Rome, to be revived when translations of Greek and Roman writings became available. During this dormant era, however, science advanced in a burst of creativity in Islam.

Within a century of Mohammed's death in 632, Arab conquerors carried their banners from Spain to the edge of China. Half a century after its founding in 762, Baghdad grew into a trade center unrivaled in wealth and importance. The conquest of Hellenistic cities, especially Antioch and Alexandria, stimulated the city's intellectual life as well. Greek manuscripts captured in such raids were translated by Jews and Christians, who were tolerated as "people of the book." Scholars became keenly interested in Greek and Roman thought, and were patronized by the more enlightened caliphs. The caliph al-Ma'mun created the famous Bayt al-Hikmah, a combination library, academy and translation bureau. He even dispatched emissaries to Constantinople in search of Greek manuscripts.

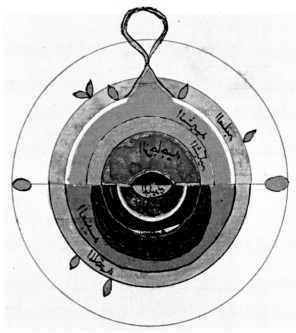

Bold colors illuminate a schematic drawing of the eye in an eighteenth-century Arab manuscript. During the Middle Ages, Islamic scientists adopted Greek learning and speculated on the nature of vision.

13

Several Arab philosophers from the ninth to eleventh centuries studied vision. Among the first was Ya'qub ibn-Ishaq al-Kindi, who deeply admired the Greek philosophers, declaring, "If they had not lived, it would have been impossible for us, despite all our zeal, during the whole of our lifetime, to assemble these principles of truth." His task, he believed, was to accurately record ancient learning, and then "complete what the Ancients have not fully expressed." In this spirit, al-Kindi authored 260 works, including a book on optics. Ears were hollow to efficiently collect sound, he observed; because eyes were both mobile and spherical, they must take a more active role. He defended Euclid's visual ray theory and attempted to prove it by eliminating those of Plato, Aristotle and Democritus.

Al-Kindi disagreed with Euclid, however, on the nature of the visual ray. Euclid had believed that the eyes projected a series of individual rays. Al-Kindi thought this "a very horrendous absurdity," since we would surely receive a spotted impression of objects. Instead, he proposed that rays flowed in single, continuous streams. Reluctant to accuse the great mathematician, al-Kindi attributed the error to followers.

Less fettered than al-Kindi by the need for geometric perfection was Abu-'Ali al-Husayn ibn-Sina, or Avicenna. Subscribing to Aristotle's view, he asserted that rays could not possibly reach the stars. Rejecting Galen's theory that eyes excited the air surrounding them, he argued, "Weak-sighted people would see better when they congregate," their collective rays boosting the air to full power.

Neither al-Kindi nor Avicenna, however, could match the achievements of abu-'Ali al-Hasan ibn-al-Haytham. Known to history as Alhazen, a scholar of the eleventh century, he was the greatest Islamic investigator of the eye. Alhazen wrote at least seventeen works on optics, integrating anatomical, mathematical and physical ideas on vision. His theories commanded respect among Europeans in the Middle Ages, eventually influencing Leonardo and scientist Johannes Kepler. Attacking the visual ray theory, Alhazen demonstrated that light affected the eye; anyone staring at the sun, he noted, felt pain. Similarly, he

Elegant geometry marks a diagram of the eye by English friar Roger Bacon. Although strongly influenced by Arab science, he advocated "crusades of learning" to impress Moslems with European knowledge.

showed that color affected the eye, for when a person shifted his gaze from a brightly colored object he saw a colored afterimage, or lingering impression of the image.

If light and color affected the eye, then objects must be the source of the rays. Alhazen therefore concluded that "from each point of every colored body, illuminated by any light, issue light and color along every straight line that can be drawn from that point." Alhazen believed the cornea, the transparent tissue coating of the eye, transmitted rays to the lens behind it. The image was then imprinted on the lens.

Medieval Light, Noble Spirit

The dazzling brilliance of Islamic creativity began to dim by the thirteenth century, hastened by the fall of Baghdad to Mongol invaders in 1258. In Europe, however, science was reawakening, as translations of Greek, Roman and Islamic writers trickled into the hands of theologians and intellectuals. Absorbing the works of Aristotle, Euclid and al-Kindi, thirteenth-century Englishman Robert Grosseteste arrived at a theory accepting visual rays. His chief interest in optics, however, lay in the spiritual nature of light. Eventually Grosseteste became Bishop of Lincoln.

A younger contemporary and fellow Oxford graduate, Roger Bacon, believed that scientific knowledge would fortify Christendom against the threat of Moslems, Tartars and even the Antichrist. In his pursuit of this goal, he studied magnetism, engineering, mathematics and geography, through which he hoped to locate the lost tribes of Israel.

Unlike Grosseteste, Bacon read translations of Alhazen. In proposing his theory of vision, however, he betrayed a characteristic weakness of medieval scholarship — the reluctance to dispute Greek dogma. Thus, Bacon attempted to synthesize several visual theories, with Alhazen cast in the leading role. Under this tidy unity, Bacon confirmed Alhazen's view that objects shed rays. But bowing to Aristotle and the visual ray disciples, Bacon added that ordinary objects lacked "nobility" and, therefore, could not produce vision. Such objects, he maintained, needed a boost from visual rays.

15

Bacon's chief accomplishment in optics was the propagation of Alhazen's ideas, which competed with rival theories until the seventeenth century. Europeans also learned of Alhazen through another thirteenth-century Englishman, John Pecham, who later in life was appointed Archbishop of Canterbury. Pecham's *Perspectiva communis* added virtually nothing to Bacon's work, but became the most popular medieval treatise on optics, running through eight printings in the sixteenth century alone.

Despite the extraordinary invention of eyeglasses in the late 1200s, the next two centuries reinforce the stereotype of the Middle Ages as an intellectually barren era. Most ancient writings had been translated. Scholars retrenched in fine-toothed analysis of established beliefs, and the theories of Aristotle dominated the curricula of Europe's universities.

Onto this stale intellectual landscape burst the Renaissance and one of its most remarkable minds, that of Leonardo da Vinci. Called the "most relentlessly curious man in history" by British art historian Kenneth Clark, Leonardo stood squarely at the crossroads of Western history. Behind him lay an intellectual tradition still dominated by the ancient Greeks. Soon to come was a scientific revolution. Leonardo's own contribution was muted, for his notebooks remained in private hands until 1636. Nevertheless, he was a key figure in the transition.

Apprenticed to a Florentine artist while still in his teens, Leonardo never attended a university. Later, he would write scornfully of his educated critics, "They strut about puffed up and pompous . . . adorned not with their own labors but by those of others." Instead, he proclaimed, he would learn through experience, "the mistress of whoever has written well." But in his anatomical studies of the eye, Leonardo would have profited from formal education. Among his many errors, he placed the lens in the center of the eye. Elsewhere, however, his lack of training worked to his advantage. He observed how the pupil shrinks and grows "according to the brightness of its object." He arrived at this insight because the phenomenon had "once deceived me in painting an eye; and that was how I learnt it."

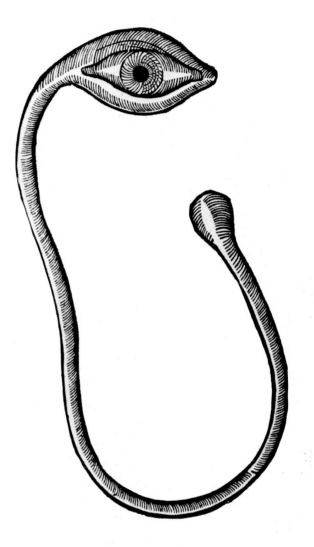

A sinuous iron wire, wrapped around the head, held a false eye against the empty socket in this device, designed by sixteenth-century French surgeon Ambroise Paré. Winning early fame for refusing to treat gunshot wounds with boiling oil, Paré chose as his motto, "I dressed him, God cured him."

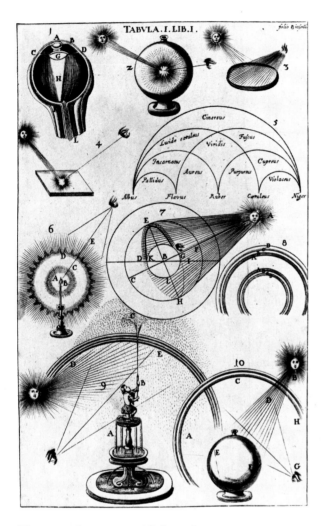

The magical commerce of light and eye was detailed by German scientist Zacharias Traber in this 1675 engraving. Drawings in the series depict the eye's anatomy, reflection of sunlight and creation of a rainbow from a fountain's spray.

Leonardo deduced that the eye operated like a camera obscura, early versions of which consisted of a pinhole in a wall admitting light to a darkened room. An inverted image of the outside scene would appear on the opposite wall. For centuries, the device was used to observe solar eclipses, and in the sixteenth century artists posed subjects beyond it and then traced their inverted images. Having discovered this fundamental principle of sight, however, Leonardo failed to recognize the crucial role of the retina. He thought that the lens reinverted the image, sending it directly to the brain.

In his earlier manuscripts, Leonardo defended the visual ray theory, referring to the power of some animals to kill, hatch eggs and attract nightingales with their eyes. Rays could reach the stars like the power of musk, which permeated "a thousand miles . . . without any diminution." In later writings he reversed himself, saying that rays "could not travel so high as the sun in a month's time."

The Death of Rays

Johannes Kepler showed no such hesitation. The brilliant son of a mercenary soldier, Kepler graduated from the University of Tübingen on a scholarship in 1591. At the age of thirty, he succeeded Tycho Brahe as Imperial Mathematician to the Holy Roman Empire. Kepler accepted the new theory of Copernicus who held that the sun was the center of the universe. Kepler himself carried the theory a step further by proposing that the planetary orbits were elliptical.

Kepler's interest in optics stemmed from his concern that vision distorted the heavenly bodies. Such distortion, he believed, would thwart accurate measurements. Like Leonardo, he recognized the eye as an elaborate camera obscura, with light entering the pinhole of the pupil. But Kepler was the first to realize that the lens focused an image on the retina. A blurred image, he asserted, resulted from improper focusing.

Kepler was deeply unsettled by his conclusion that the image, as with a camera obscura, was inverted. Yet he had faith in his own experiments. In 1625, a Jesuit priest, Christoph Scheiner, proved Kepler right when he removed

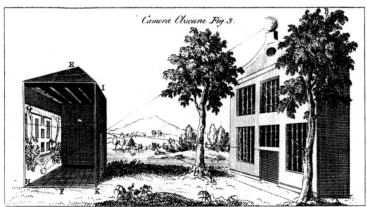

Camera Obscura Fig. 3.

Tools of the astronomer's trade fill the study of Johannes Kepler as he explains his theories of planetary motions to Holy Roman Emperor Rudolph II. Scientific lion of the seventeenth century, Kepler broke with traditional astronomy by proposing that the planets move in ellipses instead of perfect circles. In his optical studies, he deduced that the eye operates in a similar manner to the camera obscura, at left, in an eighteenth-century engraving.

Sir Isaac Newton

A Blaze of Genius

As if in defiance, young Isaac Newton stared at the full face of the sun blindingly reflected in a mirror. Its image had often filled his thoughts. Now, it haunted his eyes. After several hours in the sun's gaze, he saw the blazing orb wherever he looked. To escape it, he hid behind closed shutters in the darkness of his room for three days. Newton wrote that he "used all means to divert my imagination from the Sun." Even so, "if I thought upon him I presently saw his picture though I was in the dark."

Intellectually, Newton was rarely in the dark. Although the shy youngster grew into a somber and difficult man, Newton's genius illuminated his century. He made fresh discoveries in phenomena familiar to millions. He proved that the sun's light, thought to be the purest of all, could be split into a rainbow of colors. And in the classic tale of discovery, Newton used the sight of an apple falling to the ground to set him on the path to his theory of universal gravitation, one of the basic natural laws of the universe. "Nature to him was an open book," Albert Einstein wrote, "whose letters he could read without effort."

Newton's work with prisms, lenses and telescopes fired

his imagination and started him on a lifelong quest to understand the nature of light. The theory that grew out of his experiments became the key rival to the wave theory of light championed by Dutch physicist Christiaan Huygens.

Newton believed that light consisted of minute particles emanating in straight rays from the illuminated object. He found the idea of particles more plausible than a wave theory of light because it more easily accounted for the razor-sharp definition of a shadow's edge. Waves of light, he said, would wrap around an object, obliterating any shadow.

Newton did believe, however, that particles of light must vibrate, or undulate, and vary in size. The prism, he reasoned, divided light by the size of its particles, resulting in bands of different colors. Like

Huygens, Newton hypothesized the existence of an all-pervasive aether. But the aether, he argued, would be less dense in objects than in air. Light passed more quickly through objects than through air, he thought.

In 1704, Newton published *Opticks,* his monumental treatise on light. His theory was at first favored over Huygens's. But a century later it began to look as if Britain's most revered scientific genius had gone astray. Thomas Young, another British scientist, developed strong new evidence favoring the wave theory. About fifty years later, a new technique for measuring the velocity of light demonstrated that light slowed as it passed through objects such as glass and water, again favoring Huygens's theory.

Time, however, did not disfavor Newton. Physicists now believe that light displays characteristics of both waves and particles. Newton could not have foreseen all the scientific developments that would lead to contemporary theories of light. But this modern knowledge gives credence to the writing of English poet Alexander Pope, who honored Newton shortly after his death with the words, "God said, *Let Newton be!* and all was light."

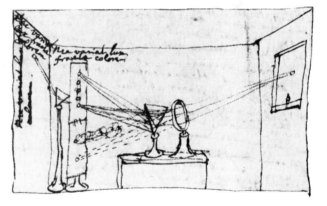

Shedding light on ancient enigmas, Isaac Newton admits sunbeams into his darkened study. Newton did not attend school until age twelve, but before reaching twenty-five he had invented calculus. In 1666, he designed a classic experiment showing that light is composed of several colors blended together. In his sketch of the experiment, below, light passes through an aperture in a window shutter, enters a prism and spreads into spectral colors.

the eye of an ox, and cut away the coating on the back. On the retina, he saw an inverted image.

If Kepler had gravely wounded the visual ray theory, Isaac Newton struck the fatal blow. Newton, a short-tempered and irrationally sensitive genius, laid the foundations of modern physics and mathematics and became the first Englishman knighted for scientific achievements.

In 1665, Newton was still an undergraduate at Cambridge, when an outbreak of plague closed the university for two years. While staying at home, he purchased a prism at a country fair. Shining light through the crystal, he observed that it refracted light. The light emerged on the opposite wall in a series of rainbow colors. The mathematical law of refraction, which determines how much light bends when it enters a transparent medium like a prism, was expressed in 1621 by Dutch mathematician Willebrord Snell. Snell failed to publish his findings, howev-

Sic quasi membrana volitant simulacra per auras
Quaq patet quouinq licet conjuncta feruntur.

A fanciful dragon, conjured by optical theorist Johann Zahn, illustrates principles of perspective in a 1702 volume. Less aesthetic, but of greater scientific consequence, is a sketch by Dutch astronomer Christiaan Huygens, whom Newton called "the most elegant mathematician" of their time. Huygens proposed that light undulates in waves, and the drawing reveals his efforts to unravel the mystery of Iceland spar, a crystal that splits light.

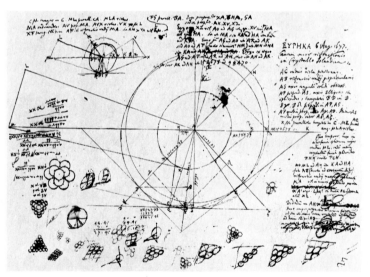

er, leaving French philosopher René Descartes to claim the credit eleven years later.

Based on his prism experiment, Newton published his revolutionary theory of light and color in 1672. He concluded that light was a concrete substance composed of particles, with different-sized particles corresponding to different colors. Together, they blended to form light. Newton further maintained that the particles moved through space in straight lines.

Provoked by criticism, he suffered a nervous breakdown and withdrew into seclusion. As Warden of the Mint, an appointment he accepted many years later, Newton vented his anger at counterfeiters and sent many to the gallows. Questions about his theory persisted, however, prompted by the publication of a rival one in 1690 by Dutch scientist Christiaan Huygens. Huygens argued that light moved in "aether," an all-pervading medium. Composed of tightly packed elastic balls, the aether carried light in waves. Newton's particles and Huygens's waves competed fiercely for scientific acceptance until the twentieth century, when it became apparent that light has properties of both particles and waves. The formulation of laws explaining light removed visual rays from scientific debate.

Christiaan Huygens

Sparring With Light

After God gave light to the world, he looked down on his creation and saw that the light was good, records the book of Genesis. Throughout history, scientists have studied the gift of light, its properties and behavior. But it took daring and pride to pose the monumental question: What is light? And it took genius to find an answer.

One man who had all these qualities was Dutch astronomer and physicist, Christiaan Huygens. Huygens's religion, he said, was science. His skill as a mathematician earned him the admiration of his illustrious contemporary Sir Isaac Newton. His discoveries in the night sky, including Saturn's rings, secured his place as one of history's great astronomers. But the specks of light that dotted the field of his telescope offered Huygens his greatest scientific challenge. They led him to explore the nature of light itself.

Huygens began his study with a comparison of light and sound. Light, like sound, he believed, traveled in waves. Sound would not travel in a vacuum, Huygens knew, but light would. Light, therefore, must be carried in a different medium than sound, something other than air and water. He believed this hypothetical medium — aether — possessed

a high degree of elasticity, allowing light impulses to ripple through its "minute, invisible corpuscles" at tremendous speed. The aether, said Huygens, was present in every substance, from air to solid bodies.

Huygens applied his wave theory of light to a unique optical phenomenon, the double refraction of light produced in a mineral called the Iceland spar. Strangely, oblique rays would pass through the crystal without refracting, but rays of light hitting the mineral perpendicularly were split in two. When Huygens looked at an object directly through the spar, the object appeared double. If he held one spar behind the first, he could see one, two or four images, depending on the way in which he rotated the crystals. Though not fully satisfied with his explanation,

Huygens attributed the phenomenon to the movement of light through both the aether within the crystal and the crystal itself.

Isaac Newton seized on this experiment as an opportunity to point out the weaknesses of Huygens's wave theory of light. The splitting of light in the Iceland spar, said Newton, indicated that a ray of light was not made up of waves. He thought a ray of light consisted of a column of particles that behaved in different ways, depending upon the angle at which it struck the crystal. Huygens's wave theory could not account for such differences, Newton argued. Both men were unaware the experiment demonstrated a phenomenon that would be recognized more than one hundred years later as the polarization of light.

Huygens's *Treatise on Light* was published in 1690, when Huygens was sixty-one. He cautioned his readers that his work was far from complete. But he concluded with confidence that some readers would find satisfaction in his book. His description of those readers could be a self-portrait; they were "those who love to know the Causes of things and who are able to admire the marvels of Light. . . ."

A distinctly unscientific challenge to Newton came from the German poet, Johann Goethe. In 1810, he published a theory claiming that color arose from mixtures of light and shadow. Scientists disdained the theory because it ignored fundamental principles of physics.

The theories of Thomas Young, on the other hand, were attacked because they challenged Newton, who by the early 1800s had achieved astounding fame. While still a medical student, Young discovered how the eye's lens changed shape for focusing. Turning his attention to light, he devised an ingenious experiment that strongly reinforced Huygens's wave theory. Young passed light through two pinholes in a screen, letting it fall on a second screen placed behind the first. Where the light from the two holes overlapped, he observed dark lines. Also a musician, Young knew that sound waves of differing frequencies could strengthen or cancel each other. The only explanation for the dark lines, he concluded, was that light waves from the different holes canceled each other as well, creating the dark lines on the screen.

Measuring the waves, Young found that different colors had different frequencies — red the shortest, violet the longest. To account for color

The dignified father of invention, German scientist Hermann von Helmholtz, below right, built the first ophthalmoscope in 1850. Within ten years, the device had revolutionized the study of the eye and its diseases. Helmholtz also created the ophthalmometer to measure the cornea's curvature. A later model, below, aided optometrists in prescribing eyeglasses.

Ophthalmometer.

CHAMBERS INKEEP OPHTHALMOMETER.

Each.

vision, he theorized that the retina harbored sensors that responded to different frequencies — and hence colors — of light. By experiment, he showed that all colors could be reproduced by combining red, green and violet. Young therefore concluded that the retina needed only three types of sensors, corresponding to these colors, to produce color vision.

Fellow scientists mocked Young for daring to question Newton. When bad publicity began to hurt his medical practice, he abandoned his investigations. French scientists pursued his ideas, however, and slowly his wave theory gained acceptance. Young's work in color vision was supported by the prestigious German scientist Hermann von Helmholtz fifty years later.

Besides confirming Young's theory, Helmholtz later made another crucial contribution to the study of vision. Realizing that a person could not peer into another's eye without blocking the light, Helmholtz built a device to solve the problem — the ophthalmoscope. For the first time, through this instrument, he saw the retina's tiny blood vessels radiating across its surface to nourish the cells that produce vision. The device was simple, its implications profound. Man could peer into the eye itself.

Halfway around the world, eye therapies figured prominently in a Japanese medical text that appeared in 1837. Below left, a patient massages his eye with a bag containing medicine wrapped around a wooden handle. A physician treats another patient, below right, with a cautery iron heated over a charcoal fire.

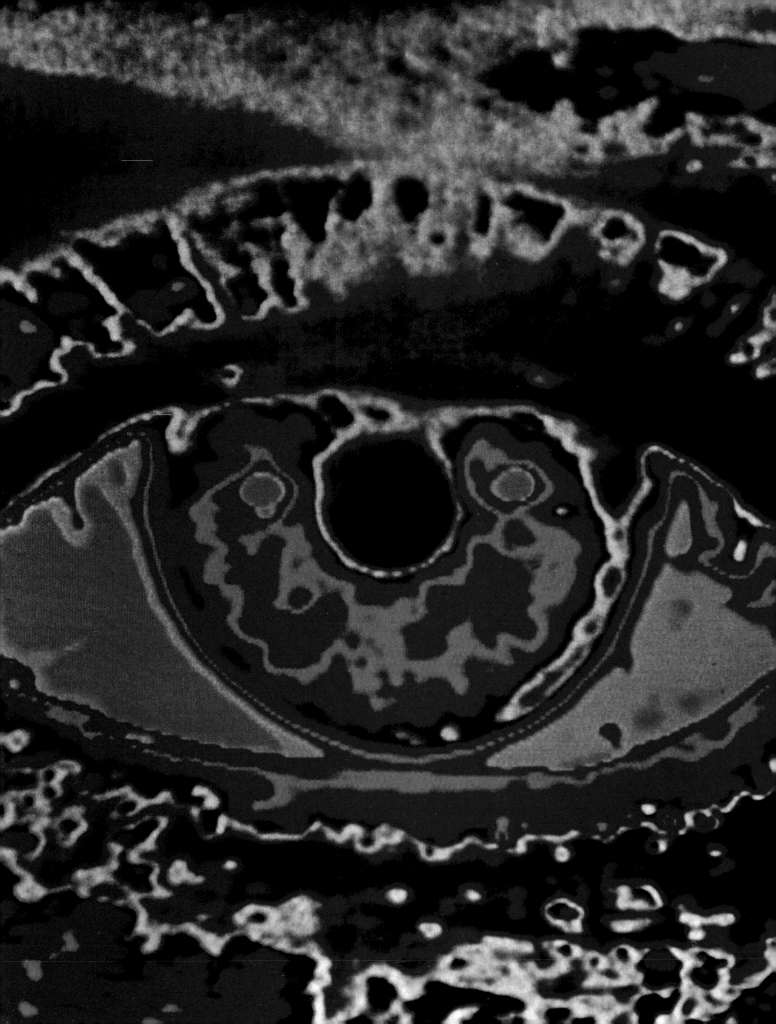

Chapter 2

The Mechanics of Vision

The twin eyes of man are simple tools. For all their enchanting beauty and power of wordless speech, they are designed to catch light. Optical devices, they gather, guide and filter light, ultimately to convert small amounts of it into the language of the brain.

Light is no less vital to man than to other creatures. The brainless amoeba flees the sun, plants grow toward it, and it is the light of springtime more than the warmth that spurs most animals to mate in the wild. While other animals strongly depend on the senses of hearing and smell, mankind puts its trust in sight. Man's sight is his principal tutor. Our eyes teach with an elegant virtuosity, guiding us through daylight with a precise tracking system that can decipher nature's subtlest signals. By night, they change into an alarm system, sensitive to shifting shadows and the dim flicker of weary light from stars uncanny distances away.

The eyes, stationed prominently — top, front and center — scout the world for the brain. While this arrangement may be the best outpost, it is also dangerous. Prey to accident, the eyes' soft tissues are cradled inside two hollow bowls of bone. Seven of the skull's bones (the frontal, lacrimal, ethmoid, zygomatic, maxilla, sphenoid and palatine) play a part in armoring and tailoring each socket. The brow ridge forms a barrier against blows and, with its fringe of hair, diverts sweat around the eyes. Lining the socket, a cushion of fat absorbs shocks and provides a well-greased surface for the constant swiveling of the eye. Stretched muscles secure each eyeball in its socket. The optic nerve, the cable linking eye to brain, also helps hold each orb in place.

When the eyes open, the central nervous system is exposed. This happens nowhere else in the body. Because of this vulnerability, the outposts of the brain are well wired for defense. A sudden movement near the face, a flash of blinding light

"Some mote, some eye-flaw, wobbles in the heat," wrote poet Robert Lowell. Revealing its novel palette, an infrared photograph maps the movement of heat through the eye. A computer assigns a different color to each shade it identifies on the infrared film. The resulting image is a way of detecting "eye-flaws" hidden behind the iris or otherwise unnoticed. Through infrared photography, the eye shines a mirror back toward its own mysteries.

27

"The tears live in an onion that should water this sorrow," wrote Shakespeare of false sadness. A subject provides biochemist William Frey with evidence that emotional and irritant tears are different.

Below, tears made in the lacrimal gland flow through the excretory ducts, bathe and moisten the eye and drain out through the lacrimal ducts. The nasolacrimal duct delivers used tears into the nasal cavity.

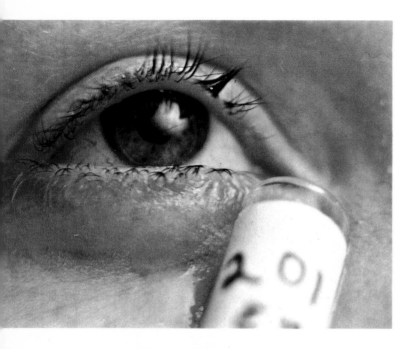

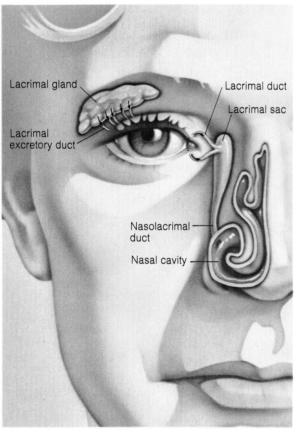

Lacrimal gland

Lacrimal duct

Lacrimal sac

Lacrimal excretory duct

Nasolacrimal duct

Nasal cavity

or a loud noise will slam the eyes shut, forming a waterproof, airtight shield. Made of tough fibers, tarsal plates shield and support the eyelids. Eyelashes are rooted in nerve cells so sensitive that a particle of grit caught by one lash will close the lids automatically.

The lids also help lubricate the eye. Sealing the eye's opening from lid to lid, and lining the inner surface of both lids, is a transparent membrane called the conjunctiva. Like a second skin, the conjunctiva snags debris that might otherwise find its way behind the orb. A thin layer of oil on its surface cuts down evaporation when the eyes are open. This membrane enables us to open our eyes under water without flooding the eye sockets. A tiny, spongelike gland, the lacrimal, supplies the conjunctiva's moisture. Resting above the eye against the socket, the lacrimal gland manufactures tears, which the lids mop over the surface of the eye. Blinking every five seconds, the lids bathe and polish the optical instrument's surface. A system of ducts behind the bottom lid drains used tears out through the nasal cavity. The conjunctiva is sensitive, and the lacrimal gland equally attentive. The pain caused by a grain of sand, the whiff of an irritating fume, even a cold wind hitting the conjunctiva, triggers the lacrimal into action. A flood of tears overflows the drains and quickly casts an intruder out onto the cheek. Stressful emotions also bring on this natural purge.

While all animals that live in air produce tears to keep the eyes moist, man is the only animal that weeps. In 1957, intrigued by the dual purpose of crying, chemist Robert Brunish analyzed the ingredients of emotional and irritant tears. Tears induced by onion fumes and strong wind, he discovered, contained a lower concentration of the protein albumin. In the 1970s, biochemist William Frey began investigating whether this protein was related to the chemical changes in our blood stream caused by stress. Tears might well play a role in filtering out the body's stressful chemicals. The machismo ethic of suppressing tears, Frey thinks, might irritate peptic ulcers and other stress-related diseases. By not allowing himself to weep, the strong, silent male might not take advantage of natural relief.

Tears also contain a mild antibacterial agent, lysozyme. Without this antiseptic, the opening of the eye would essentially be unguarded against the steady infiltration and build-up of harmful bacteria always present in the air.

The eye is designed around the retina, a screen of photoreceptor cells lining the back wall of the eye. Because the retina cannot respond to unfiltered light, optical devices of the eye stand before it, guarding the tunnel of sight like an obstacle course through which light must pass, be processed and sent on to the brain. Every part of the eye works — directly or indirectly — to deliver a clear, filtered and focused image to the retina. All parts of the eye develop from three basic layers of tissue: the sclera, an inflexible, white tunic that encases the eyeball; the choroid, the eye's blood supplier; and the retina itself.

The Living Crystal

Striking the open eye, light either bounces off the sclera, or penetrates the cornea, the crystalline bulge guarding the entrance to the eye. At the cornea, light begins its journey through the eye. The cornea slows light and bends it toward the eye's center. When this happens, all that we see before us abruptly shrinks enough to fit into a frame the size of a postage stamp. Made from the same layer of tissue as the opaque sclera, cells of the cornea are so neatly arranged that light will pass through without scattering.

Like all body tissue, the cornea must be nourished. But blood, the usual transport system, would diminish the cornea's transparency. Instead, a transparent fluid called aqueous humor feeds the cornea. Continually manufactured by specialized tissues in the choroid, aqueous humor is replenished once each hour. When the cornea's cells die, they drain along with excess aqueous fluid through a vein that circles the cornea's rim, the canal of Schlemm. The canal is so porous that even large protein molecules can easily pass through to blood vessels of the sclera, which absorb the waste. The aqueous humor's volume and pressure, critical to vision, maintain the cornea's bulge and thus its ability to bend light.

A unique feature of the cornea is the way in which its curvature decreases. Parallel rays of

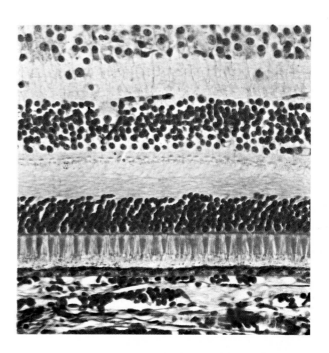

Kin to the brain, the three-tiered retina is the factory where chemistry converts light to electricity — the lingua franca of the nervous system. Light strikes the ganglion cell layer, top, and passes through the more densely packed bipolars, the middle layer, before reaching the vertically arranged photoreceptors, below. Lining the retina, the choroid, bottom, provides a nutrient bed that also traps unused light.

"And the nearest kin of the moon, the creeping cat, looked up." So penned William Butler Yeats of the luminescent globes of a cat's eyes in moonlight. What makes them glow are shiny fibers in the choroid, which reflect dim light inside the eyes, increasing sensitivity at the cost of visual clarity.

light striking the spherical cornea bend more sharply at the edges than near the center, bringing the image to an uneven focus. This distortion is called aberration. But the gradual flattening of the cornea's curve, as it meets the sclera, compensates for the distortion. Only the most expensive manmade lenses can match the cornea's precise engineering.

The sclera appears white, but its inner surface is dark tissue laden with blood vessels. This, the choroid layer, provides the main source of nutrition to the eye. But it also does more. The choroid contains a substance called melanin, the same dark pigment that colors hair and skin. By creating a dark interior, melanin traps stray light entering the eye. If light were not trapped in this way, the image would fade on the retina much as a picture on a movie screen does when the lights are turned on in a theater.

In cats, wolves and other nocturnal animals, the choroid is equipped with glistening, mirror-like fibers that volley light around inside the eyeball. Increasing by as much as 50 percent the chance of registering an image on the animal's retina, these fibers make a cat's eyes glow in the dark. But the increased sensitivity to light comes at the expense of visual clarity. Man has a greater stake in the latter.

A Flower in the Eye

Near the front of the eye, the choroid layer sloughs the roles of nourisher and absorber to take a more active part in processing an image. It forms a circular curtain of muscle called the iris, the next obstacle light must negotiate once it has passed through the aqueous humor. The black hole in the iris is the pupil. Like a thermostat constantly readjusting to maintain a desired temperature, the iris narrows and widens to maintain an ideal intensity of light passing through the pupil. Quick to respond in protecting the retina, the iris can narrow to six one-hundredths of an inch — no wider than pencil lead — allowing the eye to pick out details in the harshest glare. In the dark, it can dilate as wide as one-third of an inch to welcome any available light.

Named for the Greek goddess of the rainbow, the iris rests like a colorful flower in the eye.

Although eyes can range from steel blue to coal black, only one pigment colors them. Melanin, the same dark pigment that coats the choroid layer, determines eye color solely by the degree of its presence. Eyes appear blue or green for the same reason that the sky or sea does. When light passes through an essentially transparent medium, like the atmosphere or the aqueous humor, microscopic particles suspended in the medium scatter only light of the shortest wavelength — blue or green. Blue eyes look blue because they do not have enough melanin to color them otherwise. At birth, this pigment is hidden down in the folds of the iris, making all babies' eyes blue. Within a few months, it migrates upward, taking its place on the iris's surface to determine a child's lifelong eye color.

While the action of the iris is generally automatic, triggered by the intensity of light hitting the retina, other factors can also change pupil size. A decrease in the blood's oxygen content due to exercise can dilate the pupil. So can our emotions. Anger, fear or pleasure can be read in the eyes. For centuries merchants have known to watch the irises of their customers. Dilation betrays hidden desire, delivering a subtle message to the knowledgeable observer. That Iris was also a messenger of the Olympians seems appropriate.

The Multilayered Lens

The name *pupil* derives from the Greek word for "doll," meaning the miniature reflection seen of oneself when looking closely into another's eye. The reflection shows how dramatically the image has already been reduced to make its way to the inner eye. It seems to have depth, like a sparkling black gem, but the pupil is really the darkened surface of the eye's lens — one of nature's most unique optical instruments. The lens refines the work of the cornea to focus what we see.

When we hold a magnifying glass to an object, we move the glass back and forth to find the clearest image. The double lens system in a camera works in a similar way. One lens remains fixed while the other moves to and fro to focus. Many double lens systems in nature also work this way but, in man and other mammals, the lens remains stationary. It changes shape — its

A cross section of the human eye, dyed red, captures its mechanical simplicity. Light enters the cornea, the fluid-filled bulge at right, bends sharply toward the center and meets the spherical lens. Despite its size, the lens makes only minor focusing adjustments for light's journey to the retina, the dark, fine line that circles the interior back wall of the eye. Behind the retina, the choroid traps stray light, while toward the front of the eye, it sprouts the iris, which works to maintain an ideal amount of light entering the eye. Messenger of vision, the optic nerve, at left, delivers sight to the brain.

Optical fibers of the human lens, when magnified more than 800 times, look like meticulously stacked lumber. The multilayered arrangement yields precision to the eye's task of focusing by bending light smoothly and gradually before it strikes the retina.

thickness and surface curvature — to sharpen the picture produced by the cornea. This process is called accommodation.

Like an onion, the human lens is built of layers of cells. Neatly arranged, the layers allow light to pass through without scattering. Each layer bends light a bit differently, but the overall effect is smooth and gradual. Seventy fingerlike ligaments collectively known as the ciliary zonule hold the lens in place. The fibrous ligaments, in turn, are attached to the choroid coat by the circular ciliary muscle. When the zonule pulls the perimeter of the lens, the lens stretches into a flattened shape. This posture, the resting position of the eye, requires no adjustment to bring objects more than twenty feet away into focus. For the eye to focus an object within twenty feet, the ciliary muscle contracts, relaxing the zonule's pull and allowing the lens to bulge. The nearer an object, the more the lens assumes its rounded, balloon shape. In the human eye, accommodation has a limit. Within six or seven inches, an object reaches the "near point," beyond which the lens cannot accommodate enough to focus. Many birds and reptiles have stronger ciliary muscles that squeeze the lens, making it pear-shaped to allow an almost microscopic acuity.

The lens grows throughout life. As more and more layers accumulate, older ones pack together toward the center, eventually forming a hardened core. In proportion to the hardening, the lens loses its flexibility, making the near point recede. By middle age, accommodation stops altogether; the focusing system of cornea and lens locks onto a specific distant point. To see anything within or beyond this range, the eyes require help, which glasses provide. As a physician and man of letters, Oliver Wendell Holmes took delight in observing people. Watching a man reading, Holmes noted whether it was necessary for the man to move the pages back and forth. If it were, Holmes could deduce the fellow had reached the "trombone age," between forty and fifty.

The lens also acts as a filter. Slightly tinted yellow, it blocks out all light beyond the violet end of the visible spectrum. Cataract patients whose lenses have been surgically removed and replaced with untinted glass see clearly in ultra-

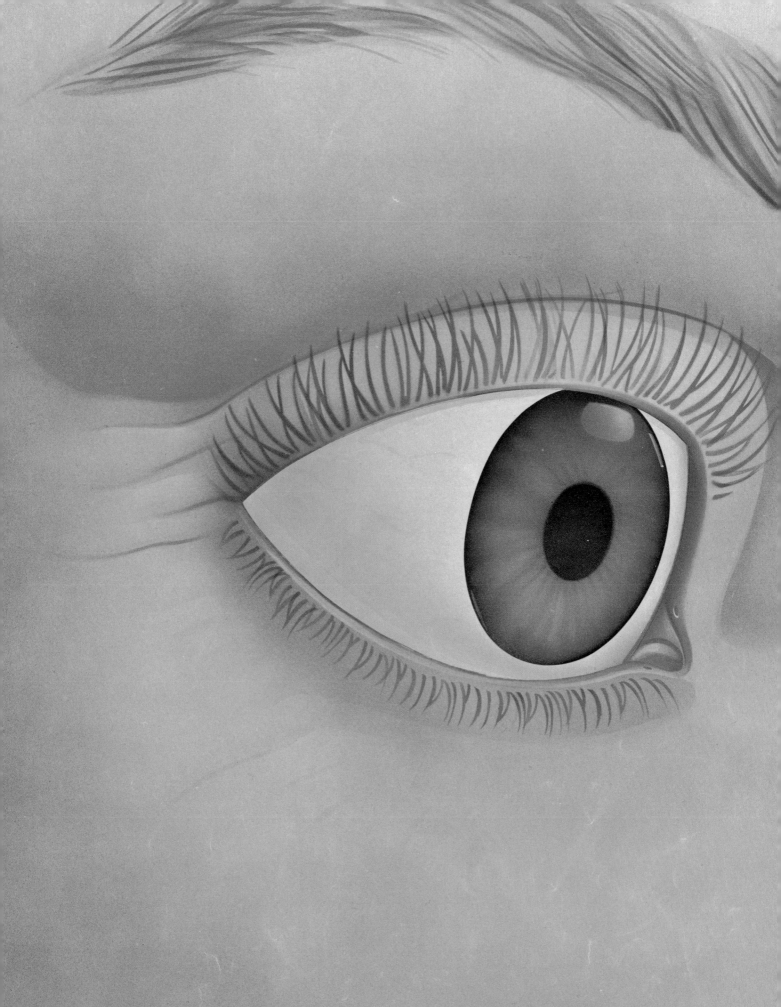

Superior oblique muscle

Superior rectus muscle

Frontal bone

al rectus muscle

Lateral rectus muscle

Conjunctiva

erior rectus muscle

Inferior oblique muscle

Punctum lacrimale

Medial canthus

Maxilla

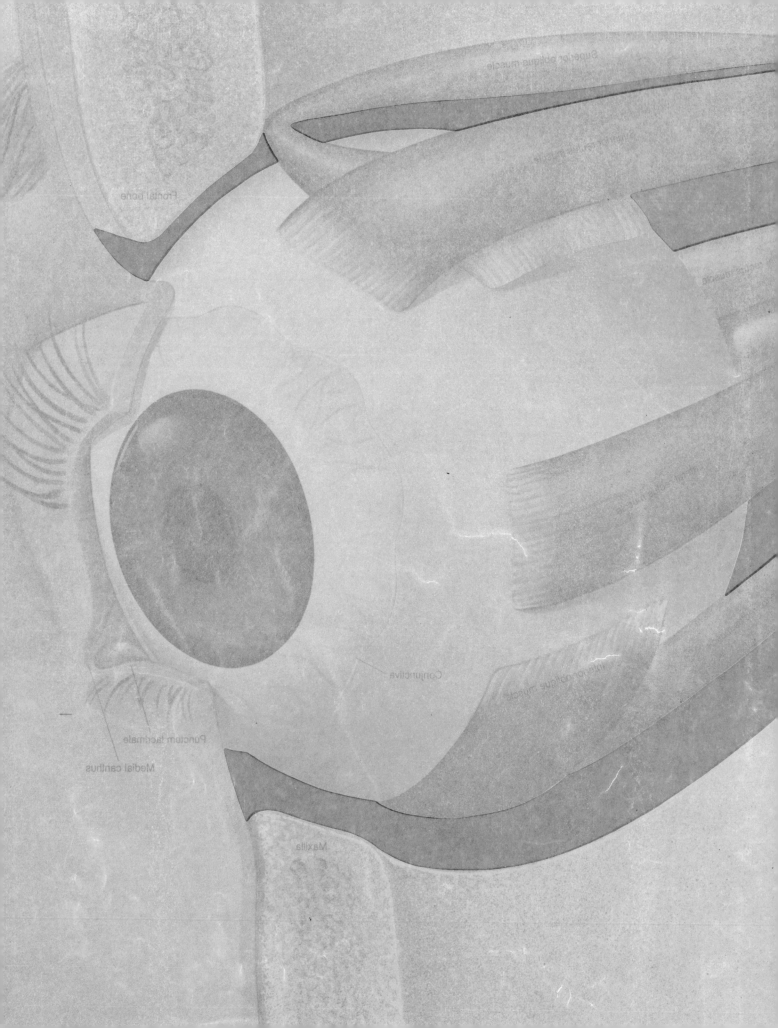

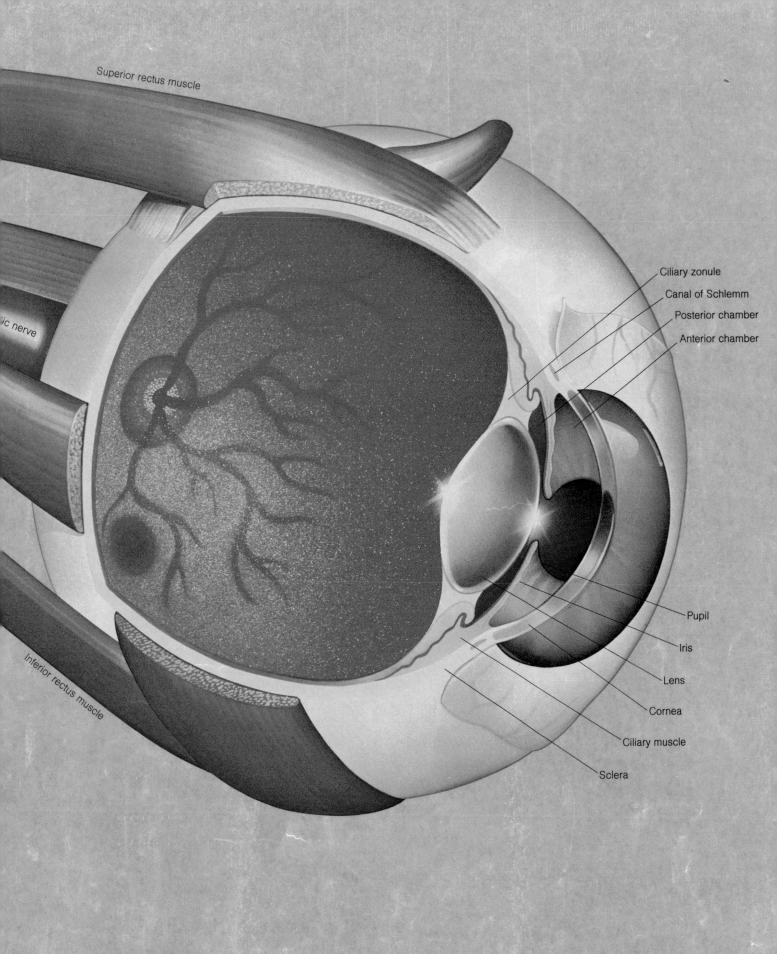

Superior rectus muscle

ic nerve

Inferior rectus muscle

Ciliary zonule

Canal of Schlemm

Posterior chamber

Anterior chamber

Pupil

Iris

Lens

Cornea

Ciliary muscle

Sclera

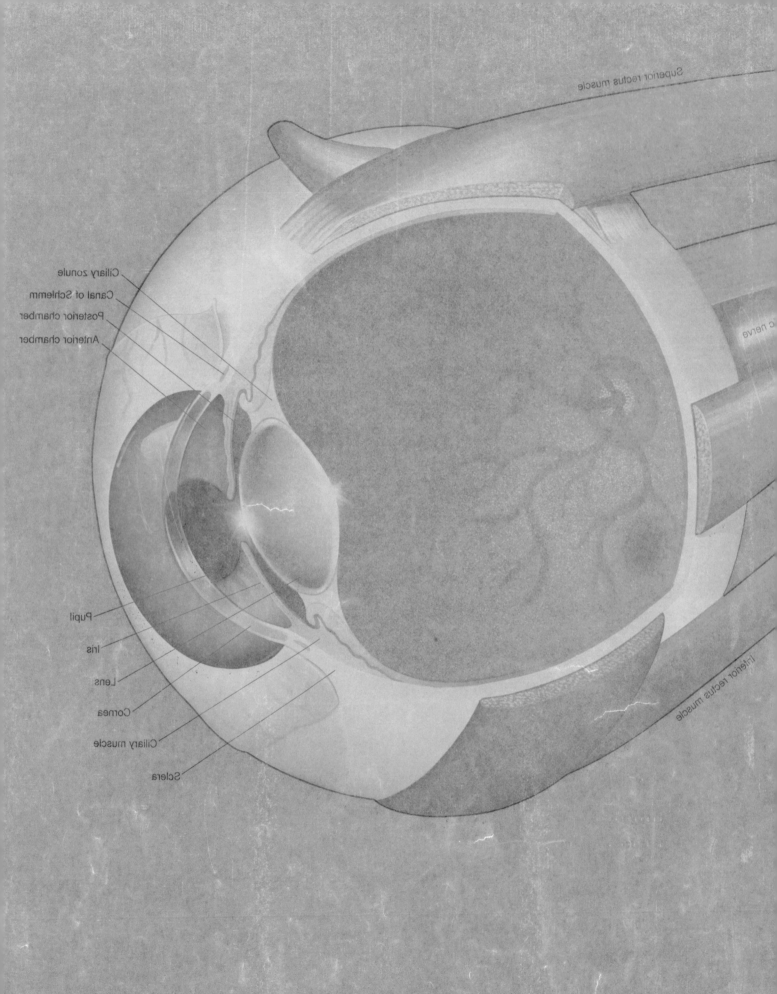

Superior rectus muscle

Ciliary zonule
Canal of Schlemm
Posterior chamber
Anterior chamber

c nerve

Pupil

Iris

Lens

Cornea

Ciliary muscle

Sclera

Inferior rectus muscle

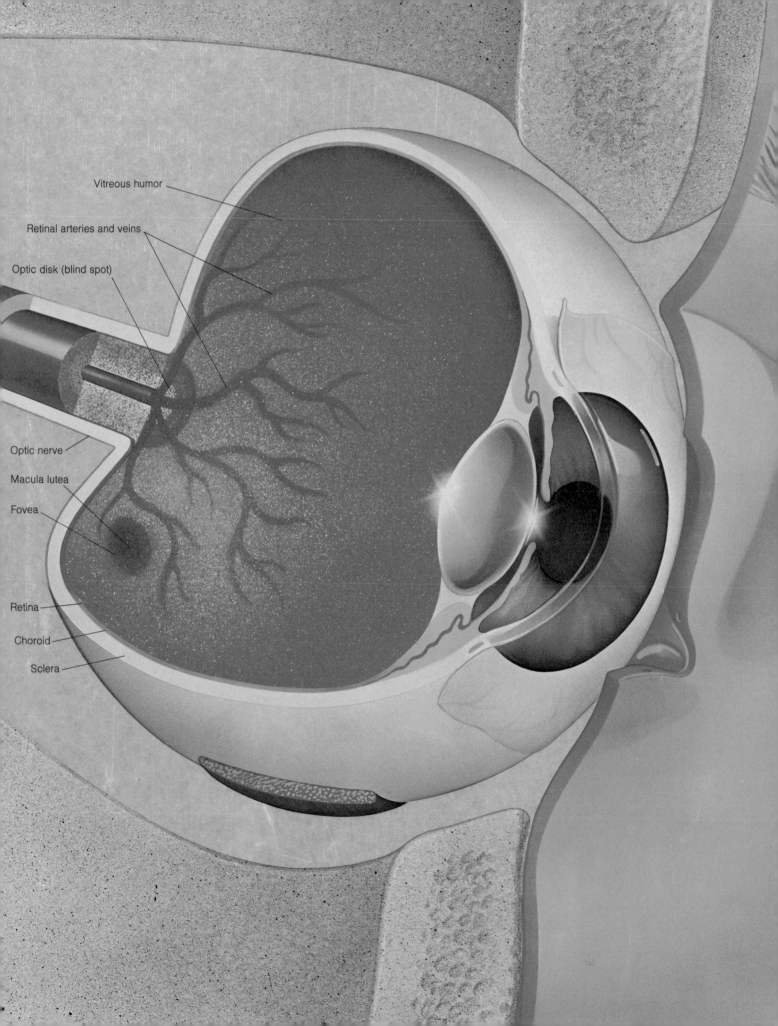

Vitreous humor

Retinal arteries and veins

Optic disk (blind spot)

Optic nerve

Macula lutea

Fovea

Retina

Choroid

Sclera

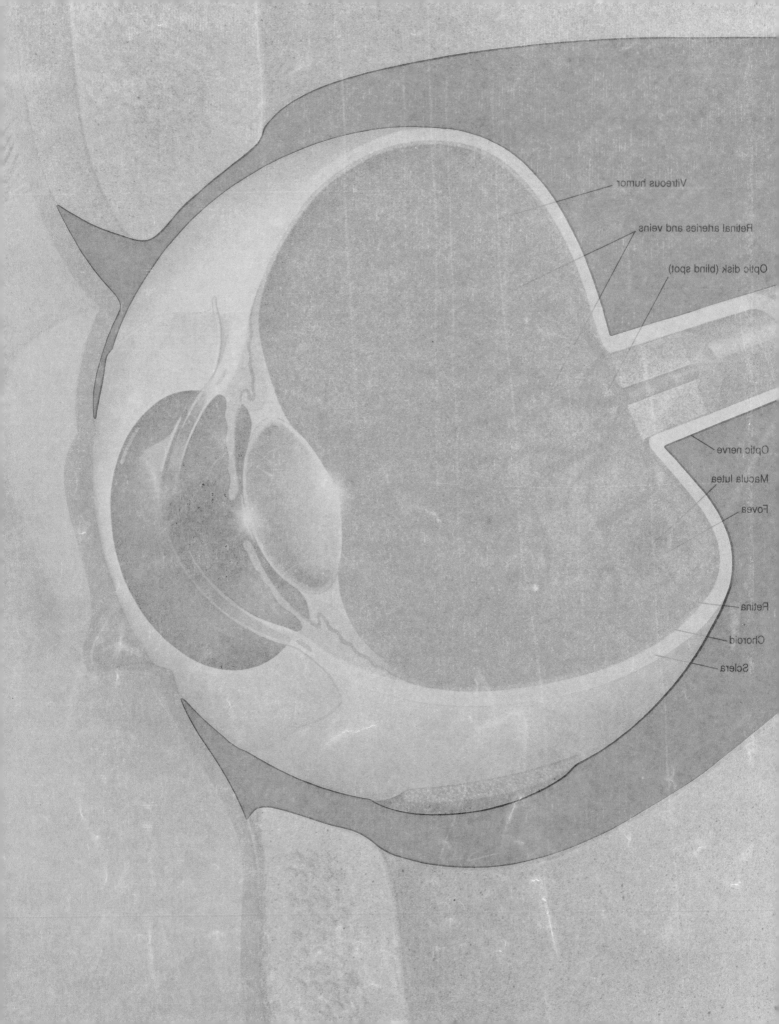

Johannes Kepler

The Geometry of Vision

On July 10, 1600, Johannes Kepler stood in the center of a bustling marketplace peering into an odd-looking box. A small image appeared on the darkened wall of the box, admitted by a pinhole opening on the opposite side. It showed the moon creeping across the face of the sun, blocking its light. Kepler was so engrossed in his work that he did not notice the pickpocket reaching for his purse. The thief made off with thirty guldens. But the German astronomer, who would later formulate the laws of planetary motion, was on his way to riches of another kind.

The blackened box, called a camera obscura, had been known for at least 600 years before Kepler's time. Light beaming through the pinhole opening produced an inverted but proportionally correct image of an object on the box's back wall. Intrigued by the image, Kepler decided to investigate the effect of a small aperture on rays of light.

Kepler used an elevated book to represent a light source and tied strings to its four corners and sides to represent light rays. He drew the strings down from the book through a small opening in a table and onto the

floor. By the time the strings hit the floor, they had crossed; yet, in reverse, they maintained the shape of the book. Kepler believed light rays crossed the same way in the camera obscura.

But what of the eye? Kepler wondered. The box inverted its images, he knew, so perhaps the eye did, too. Several of Kepler's predecessors had noticed similarities between the eye and the camera obscura. But earlier scientists did not believe that the eye inverted images. If it did, they reasoned, the world would appear upside-down. The riddle of the inverted

image troubled Kepler, too. Yet, his painstakingly precise geometry of rays forced him to accept what other great minds had found incredible: The human eye saw a world upside-down.

Kepler combined his studies on the geometry of rays with his knowledge of the anatomy of the eye to probe the eye's inner workings. He believed that rays of light entering the eye passed through the lens, known in Kepler's time as the crystalline humor, and focused into cones of light directed at the back of the eye. The rays converged on the eye's smooth, back wall, said Kepler. There — not on the lens, as others believed — the image formed. "Therefore vision occurs through a picture of the visible thing on this . . . concave surface of the retina," Kepler concluded.

The problem of the inverted image remained unexplained, but Kepler was content to "leave it to the natural philosophers." He sensed that the process was one more of mind than mechanics. But his poetic imagination found a whimsical explanation for this riddle of sight. Kepler envisioned the strange, inverted images on the retina met at the back of the eye by "a magistrate sent by the soul."

violet, or "black" light. With age, the yellow tint darkens, encroaching on the violet and blue colors of normal vision. That some painters, late in their careers, shy away from purples and blues has been attributed to this darkening of the lens.

Emerging from the back of the lens, an image has been reduced, dimmed, filtered and focused. Yet it has only traveled one-third of its journey to the retina. It must still pass through the largest portion of the eye, a ball of clear jelly called the vitreous humor. This gelatinous network keeps the eyeball under light pressure, maintaining its spherical shape so that it swivels smoothly. Dead blood cells occasionally invade the vitreous. Casting vague shadows before the eyes, they are the tiny spots — called *muscae volitantes*, or "flying flies" — that float away when we try to follow their path. The vitreous is generally as clear as water, permitting the image to pass through the eye, undisturbed, to the retina.

The Brain's Photon Net

Ancient scholars named the *retina* — meaning "net" — for its intricate crisscrossing of blood vessels. Some reasoned that the retina nourished the eye. Others, more fancifully, imagined it a light-beam generator. They would never know how close they had come to the crux of vision. For the retina is our catcher of light, the brain's photon net. A thin, red membrane, this interior layer of the eye contains, in turn, three distinct microscopic layers of nerve cells. The optic nerve, entering the back of the eyeball, unravels into a million fibers that plug into the innermost layer of ganglion cells. Each ganglion attaches to one or more cells in the middle layer, the bipolars. These are transfer cells, passing information from the outermost layer of photoreceptors. The miracle of translating light into electricity — the language of the brain —takes place in this, the final microscopic layer of the retina.

Throughout the animal kingdom there are but two types of photoreceptor cells. Cones give sight by day and rods, by night. Elongated, tube-shaped cells, the photoreceptors stand like a densely packed forest rooted in an earth of bipolar cells. Their light-sensitive tips protrude from the retina's membrane to form a mosaic screen.

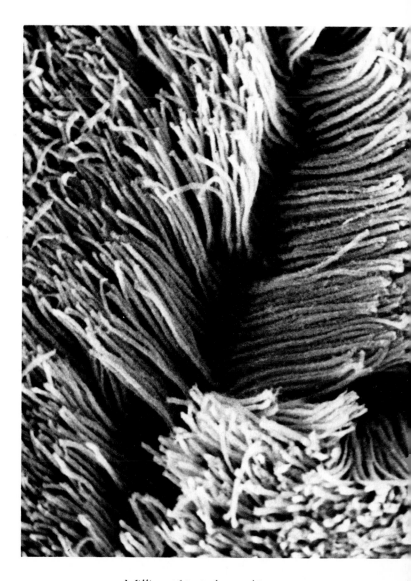

Millions of tentacles reaching out for light, photoreceptors called rods line the retina. About 120 million per eye, they contain light-sensitive pigments that produce peripheral and night vision.

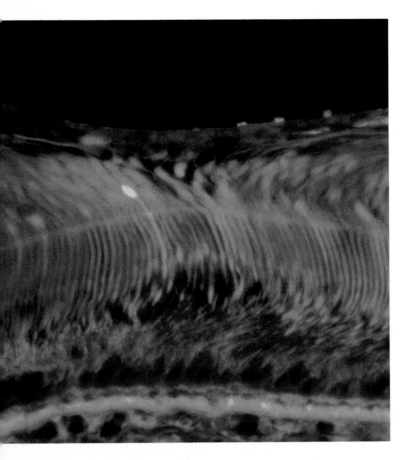

The fovea, or "pit," of the retina provides extra acuity to our vision. Directly aligned with the center of the pupil, ganglion and bipolar cells move aside to form this indentation and allow light its only unblocked path to photoreceptor cells. Specialized cone cells — smaller and more tightly packed than usual — wait at the bottom of the pit to receive the unobstructed light.

The shapes of the photosensitive tips distinguish cones from rods; but the differences go beyond form. Rods react to as little as one photon, the smallest unit of light. Rods send the brain only gray, hazy images. Cones are a thousand times less sensitive but, with enough light, they "see" the world's colors and fine detail. Cloaked in dark pigment, the photoreceptors of the eye face away from light. An image must pass through the ganglion and bipolar cells before reaching the photoreceptors, somewhat like projecting a picture through the back of a screen.

Following the general law that nature's gifts are tailored to their owner's needs, rods and cones are distributed differently among animals. The high-flying kestrel has a rich store of cones — 600 million closely packed in a square inch of retina's center —giving this bird of prey a visual acuity eight times that of man. The owl, a night hunter, can see ten times better than man in the dark, by virtue of its more numerous rods. The dogfish and eel, which feed and spawn on the murky ocean floor, have retinas made only of rods. The chicken has only cones; while it might be able to spot the smallest insects in barnyard dirt, at night, it is totally blind and helpless against the invading fox.

The human retina has almost eighteen times more rods than cones; but their arrangement on the screen makes our vision a formidable tool both day and night. In the center of the retina, directly in line with the path light follows through the pupil, lies a yellow spot called the macula lutea. This area, one-tenth of an inch in diameter, forms a shallow crater rich in small cones bunched tightly together. The center of the macula, the fovea, is a "pit" one-fourth the size of the macula. Here, the fabric of the retina, through which light must normally pass to reach the photoreceptors, has been brushed aside. It is the only place on the retina where light has un-obstructed access to the photoreceptors. Its task is to focus what we see. To look at something is to cast its image on the fovea.

Alongside the macula, nerve fibers gather to form the optic nerve. Their pathway through the screen makes a hole where no vision occurs. This "blind spot" is easy to find: write a small *x* and

Human vision is dual, the work of two types of photoreceptors. Cones detect color and fine detail, yielding daylight vision; while rods provide the eyes with the ability to see hazy images in fading twilight.

another, about three inches from the first, on a piece of paper. Keeping the right eye closed, focus the left eye on the right x and move the paper slowly toward the eye. About ten inches from the eye, the x on the left will vanish as its image hits the blind spot. This hole in the canvas of vision is always with us, but the brain has learned to fill in what the eye has missed.

The farther away from the fovea, the fewer the cones. At the edges of the retina, the sparse cones are virtually lost in a forest of rods. The periphery of vision is the domain of the sensitive night receptors. That is why dim stars are only visible when we look next to — but not directly at — them; why, on the edges of our field of vision, we see only in black and white; and why we can detect something moving in the corner of our eye without being able to identify what it is.

Daylight vision (photopic) and night vision (scotopic) gear the human eye to the tempo of

life on earth. They phase in and out comfortably with the rising and setting of the sun. The process is known as adaptation. Rushing the transformation temporarily breaks down the visual system. Quickly moving from a darkened room into daylight can be not only blinding, but painful. Together, daylight and night vision stretch our sensitivity to light.

Vision's Chemistry

Between darkness so black that we cannot see and brightness so glaring that it blinds us, the human eye's range of sensitivity is awesome — from the glowing tip of a cigarette in the dark to a white paper in the noonday sun, a surface one million times brighter. Using controlled flashes of light on a darkened screen, researchers have been able to plot the process of adaptation. Introduced to darkness, the light-adapted eye increases its sensitivity at a rapid rate. In the first minute, the eye is 10 times more sensitive; in twenty minutes, 6,000 times more sensitive; and after forty minutes, the eye reaches its limit. When it is fully dark-adapted, 30,000 times less light is enough to activate the retina. The iris, too, plays a role in adapting the eye to light by enlarging or narrowing the pupil. But its role is minimal. When the pupil is completely opened, the eye is only thirty times more sensitive to light than when the pupil is at its smallest. The greatest adaptation to light and dark is accomplished in the retina itself.

In 1876, working in a darkened laboratory, German physiologist Franz Boll removed the retina from a frog's eye. Moments after exposing it to light, Boll watched the bright red tissue fade to orange and soon to bleached gold. With live frogs, he found that a strong light would also bleach the retina; but in the dark, over time, the living tissue would again become red. He was convinced this process of bleaching and regeneration, a reaction to light and dark, was responsible for vision. Boll called the mysterious red pigment *erythropsin,* or "visual red." Although he failed to discover the chemicals in the process, his hunch about the mechanism of bleaching and regeneration would later prove accurate.

Within a year, his countryman, Wilhelm Kühne, found the pigment in the retinas of other animals, including mammals. Because it was darker than the frog's red pigment, he renamed it *rhodopsin,* meaning "visual purple," the term scientists still use. Kühne and his assistant extracted the chemical, dissolved it in salt water and studied its behavior in light. In two prolific years, they documented everything that was to be known about rhodopsin for the next fifty years. What the chemical was and how it worked remained nature's secrets, until American biologist George Wald took up the search in 1932.

As a graduate student at Columbia University, Wald worked in the laboratory of Selig Hecht, "one of the great measurers of vision," Wald later recalled. Hecht postulated that light broke rhodopsin into two compounds, stimulating the photoreceptors. Through some normal body function, he reasoned, the two compounds must rejoin to form rhodopsin. Once this was accom-

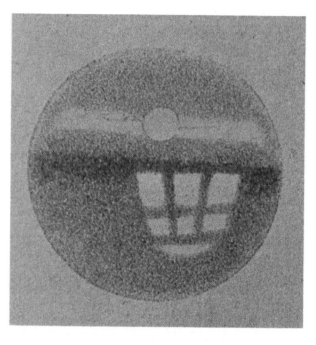

plished, the cycle repeated itself in living tissue. Hecht translated the cycle into a simple equation, with S representing rhodopsin; and A and B, its two unknown components: $S \rightleftarrows A+B$

"I left Hecht's laboratory," Wald later wrote, "with a great desire to lay hands on the molecules for which these were the symbols." In 1933, working in the laboratory of Otto Warburg in Berlin-Dahlem, Wald discovered vitamin A in the retina. The concentration of the vitamin increased as rhodopsin bleached and decreased as the pigment returned. Wald had found one of the components of rhodopsin.

Knowing the body did not manufacture its own vitamin A, Wald realized a diet lacking it should greatly affect vision. He sought and found historical evidence showing that man had been aware of the connection between diet and sight for more than 4,000 years. Without knowing the ingredient, physicians through the ages had recommended foods containing vitamin A to patients suffering from night blindness. By trial and error, perhaps, ancient Egyptian physicians had prescribed the raw liver of donkey and black rooster to patients whose vision failed in dim light. Pliny the Elder, great Roman naturalist of the first century A.D., had suggested goat liver

Wilhelm Kühne

Portrait in a Dead Man's Eye

Reading by the light of a solitary lamp, the man flicks the pages of a novel, unaware of the shadow that creeps along the baseboards of the paneled room. Suddenly, there is a sound. He turns, his eyes taking in the image of a man firing a gun. Later that night, a detective entering the dark room does something strange. He takes a picture of one of the dead man's open eyes, hoping to catch the image of the man's killer still lingering on the victim's retina.

The notion that the eyes of a dead man could contain the portrait of his murderer was a misconception that arose from the work of nineteenth-century German physiologist Wilhelm Kühne. It was a notion Kühne despised. "Indeed it is no pleasant thing to find a serious study considered as a fit companion for such ideas," he wrote. Kühne's work was not aimed at helping detectives. He sought to unlock one of the secrets of vision.

In 1876, German biologist Franz Boll discovered that a pigment in the retina could be bleached in the sunlight but returned to its original color in the dark. He believed the pigment, later known as visual purple, or rhodopsin, was part of the photochemical process that prompted visual impulses. Kühne succeeded in confirming Boll's findings and, almost

accidentally, developed a process he called optography. The eye of a rabbit became Kühne's camera, the retina his photographic film.

Kühne placed a rabbit in a dark room and fixed its gaze on the only source of light, a barred window. He covered the animal's eyes with a black cloth for ten minutes, then withdrew the cloth and for two minutes exposed its unblinking eyes to the barred window. Next, Kühne swiftly decapitated the rabbit, keeping its head covered, and removed one eye. He sliced off the portion containing the retina and placed it in an alum solution, which dried the retina and acted as a photographic fixative to preserve the image. "On the third day afterwards," Kühne reported, "I had the pleasure of recognizing in the removed retina the image

of the window to which it had been exposed, its arched ends appearing as white silhouettes on a red ground, and between them several smaller clear fields."

Kühne successfully repeated the experiment with the rabbit, later trying to capture more complicated subjects on the retina, including a garden and a portrait of a man, but found "both as yet leave much to be desired." He was reluctant to extend his conclusions to include human eyes until he could examine them similarly.

His opportunity arrived in November 1880, when a criminal was beheaded in a nearby town. Kühne brought the body into a dimly lit room, cut the retina from the left eye and beheld an enigmatic shape strongly bleached on its surface, a rectangular figure with jagged steplike markings on one side. Kühne retraced the man's last hours, interviewed witnesses and studied the scene for any clues to the origin of the mysterious shape.

Although he failed to discover the optogram's source, his long scientific search confirmed the striking similarities between eye and camera. Kühne proved that light brands the surface of the retina, like the film of a camera, with images of the outside world.

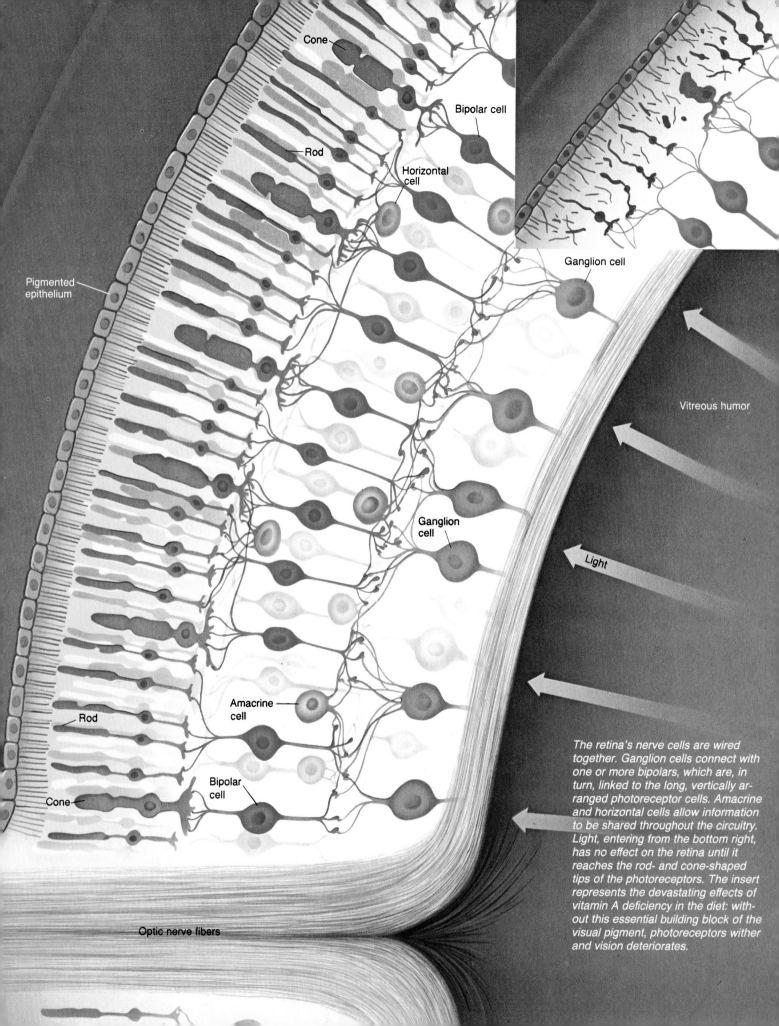

Cone

Bipolar cell

Rod

Horizontal cell

Ganglion cell

Pigmented epithelium

Vitreous humor

Ganglion cell

Light

Amacrine cell

Rod

Bipolar cell

Cone

The retina's nerve cells are wired together. Ganglion cells connect with one or more bipolars, which are, in turn, linked to the long, vertically ar-ranged photoreceptor cells. Amacrine and horizontal cells allow information to be shared throughout the circuitry. Light, entering from the bottom right, has no effect on the retina until it reaches the rod- and cone-shaped tips of the photoreceptors. The insert represents the devastating effects of vitamin A deficiency in the diet: with-out this essential building block of the visual pigment, photoreceptors wither and vision deteriorates.

Optic nerve fibers

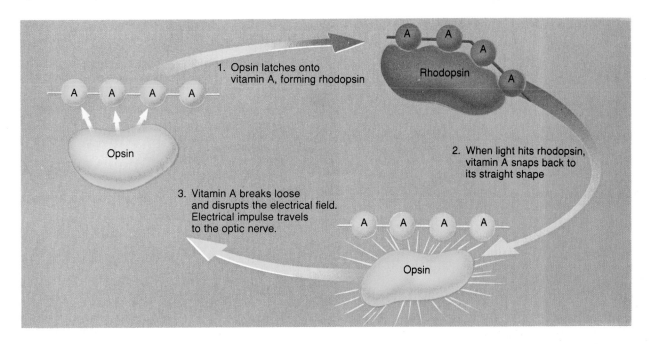

1. Opsin latches onto vitamin A, forming rhodopsin

Rhodopsin

2. When light hits rhodopsin, vitamin A snaps back to its straight shape

3. Vitamin A breaks loose and disrupts the electrical field. Electrical impulse travels to the optic nerve.

Opsin

Opsin

since it was said that goats saw "no less well at night than in the daytime." Cod-liver oil became a well-known remedy in modern times. But it was not until 1917 that vitamin A was discovered to be the active ingredient in liver's curative power. The clues, Wald realized, had been there all along, laid out in a logical path that could have led anyone to his discovery. Although his own route was more arduous, Wald had pieced together a centuries-old puzzle.

His discovery was not only a landmark in eye research, but also the first study proving that a vitamin played a role in normal body functions. In accepting the Nobel Prize in 1967, Wald called his work "a quiet conversation with Nature. One asks a question and gets an answer. . . . an experiment is a device to make Nature speak intelligibly. After that one has only to listen."

Wald's "conversation" led him to a model for the chemistry of vision: when the vitamin A molecule enters the body (in leafy green vegetables or liver), its shape is long and straight. It migrates to the choroid, where photoreceptors can absorb it. Entering a rod, the vitamin meets a colorless protein molecule called opsin. Opsin latches onto the vitamin, bends it into a hooked shape and forms rhodopsin. This compound

spreads throughout the light-sensitive tip of the cell. When one photon — the smallest unit of light — hits a rhodopsin molecule, the pigment explodes. Vitamin A snaps back to its straight shape and breaks loose from the opsin, disrupting the electrical field of the cell. Bipolar and ganglion cells then rush the message of vision to the brain. By 1959, Wald's colleagues at Harvard were ready to declare that the only service light performs is to change the shape of the vitamin A molecule, releasing it from the pigment complex. "Everything else," they wrote, "further chemical changes, nerve excitation, perception of light, behavioral responses — are consequences of this single photochemical act."

Rhodopsin is found only in rods, but Wald's model also proved workable for explaining the chemistry of color vision. In cones, vitamin A combines with three different opsins, creating three pigments, one for each of the three primary colors — red, green and blue. These pigments are much more stable and selective than rhodopsin — they require certain wavelengths of light to explode. Each cone, depending on which pigment is most abundant, specializes for one of the three primary colors. Combined, the cones provide the brain with a palette that can mix any color.

41

The high concentration of rhodopsin
in the alligator's dark-adapted eye
heightens sensitivity and gives its
interior a rose-colored appearance.
The photographer has captured the
color before adaptation begins.

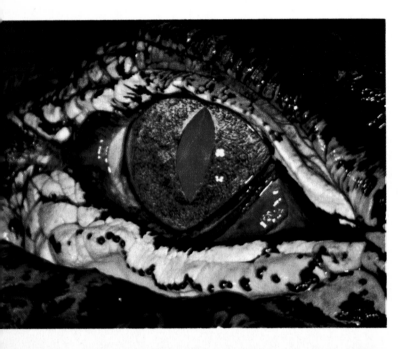

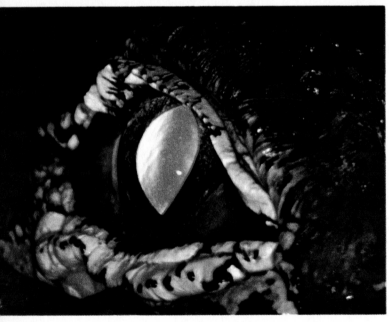

*After moments in the light, the
concentration of rhodopsin drops and
the inner eye appears pale gold. This
cycle of bleaching in light and
regenerating in darkness was the
first clue to the chemistry of vision.*

Was the eyes' sensitivity in darkness attributed to the amount of rhodopsin present? Wald's chemistry provided a strong argument that it was. In bright light, rhodopsin breaks down faster than it regenerates, so much faster that the rod's concentration of fuel drops low enough to shut down the cell. As light intensity fades with the setting sun, rhodopsin begins to regenerate faster than it bleaches, and the rods power up to begin operation. There is a point in the process of adaptation when rods and cones work together, enabling us to see the faint, grainy colors of dusk. But eventually, without enough light to stimulate them, the cones cease to work. After forty minutes of darkness, the rods have regenerated the maximum concentration of rhodopsin to reach their height of sensitivity.

Researchers in Wald's laboratory roughly estimated the number of rhodopsin molecules in the dark-adapted eye (10 million in each rod; 100 million rods in each retina). On the assumption that one photon of light broke apart one molecule, they presented measured amounts of light to a fully dark-adapted eye. The decrease in the eye's sensitivity matched the amount of light absorbed. The dynamics of the chemical equation fit the rate of adaptation. From such neatness emerged a photochemical theory of vision.

In the 1950s, British physiologist W. A. H. Rushton devised an ingenious experiment to prove there was more to changes in the eye's sensitivity than just the concentration of pigments. He found a way to shoot minute parcels of measured light into the dark-adapted eye. With a light meter, he caught the reflected light, and, so, could estimate the number of photons that had not caused any photochemical changes. When the concentration of pigments dropped only slightly, the eye's loss of sensitivity was, by the chemical equation, inordinately high. There must be more at work in the eye's sensitivity than chemicals, Rushton argued.

His suspicion echoed that of Gustav Theodor Fechner, who, a century earlier, had laid the groundwork for the measurement of vision in his classic *Elements of Psychophysics*. Fechner projected a minute beam of light into a subject's dark-adapted eye. As expected, the iris narrowed. But

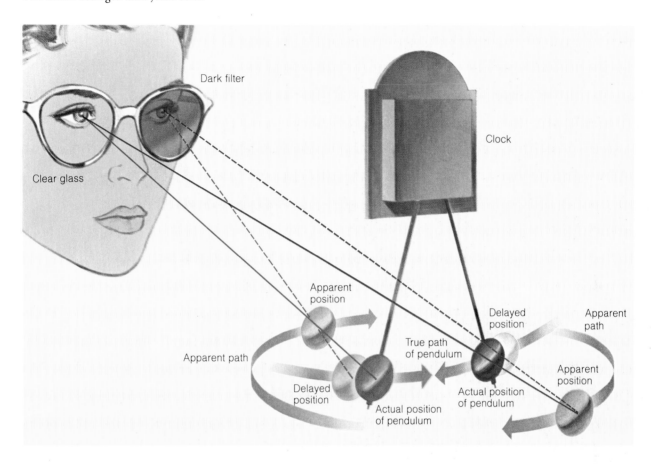

when Fechner aimed a second, smaller beam at the eye, something unexpected happened. Since the total amount of light hitting the eye had been increased, he expected the iris to narrow even more. Instead, the iris widened, adjusting to the average amount of light rather than the total. His experiment is still known as the Fechner paradox.

There is evidence that the wiring of the retina might allow the rods and cones to share information. In the late 1960s, John Dowling of Johns Hopkins studied the retina's bipolar and ganglion layers and found that these nerve cells were heavily interconnected. Special horizontal cells linked the bipolars together, while amacrine cells connected the ganglions. Dowling suggested that the information from light did not travel a one-way route to the brain, but might also be relayed within the retina itself. Repeating Rushton's experiment, researchers have found that although only a small amount of light strikes the photore-

ceptors, its presence immediately becomes general knowledge throughout the retina. The eye begins the process of adaptation at once. In the Fechner paradox, some scientists have speculated that the two beam intensities could be averaged within the retina before going to the brain, which adjusts the iris accordingly. Just how the retina accomplishes this task remains a mystery, but it is certain that chemistry is only one of the steps.

The retina's photoreceptors also aid in the detection of motion. In an experiment called the Pulfrich pendulum effect, an observer puts on a pair of sunglasses with only one lens. While one eye adapts to dark, the other remains light-adapted. Watching a pendulum swinging from right to left, the subject sees the pendulum swing in a broad, oval motion even though he might know that its course swings along a straight line. The illusion created is the work of the photoreceptors in the dark-adapted eye. Its rods respond

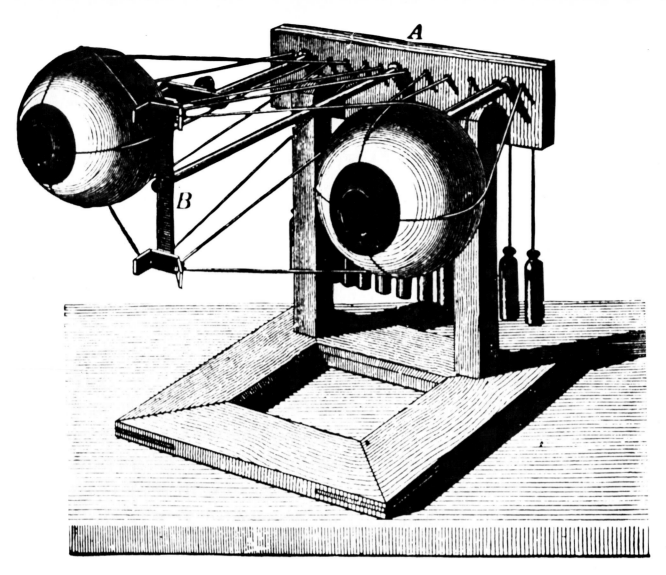

more sluggishly than the cones in the light-adapted eye. The reason lies in the wiring of the retina. Cones have more exclusive electrical connections to the optic nerve, like a hot line to the brain. In the fovea, each cone has its own bipolar and ganglion cell. Rods, however, share party lines; a minimum of five rods must simultaneously fire to complete a call. Even then, it is difficult to trace which cell initiated the call. The dark-adapted eye sees a delayed image of the ball. The brain averages the information from both eyes, but is completely fooled.

The Pulfrich pendulum effect demonstrates two of the most important aspects of vision: the accurate perception of motion and the necessary coordination of both eyes to bring it about. In the drama of survival, stationary objects are just props — neither threat nor prey. When objects move, the brain must be alerted to command an appropriate reaction.

Some movements are too fast, too slow or too distant to be perceived. A snake can stalk its prey so slowly, and strike so swiftly, that neither movement can be detected by the human eye. While we can say with the passing of time that the snake is in a new position, we cannot see it move. Should the snake blend in with its background, we, and the snake's potential dinner, might not see it at all. Size, the amount of light, the motion of nearby objects all affect the eye's sensing of motion.

The eyes sense motion in two ways. When the observer's eye remains still, the image of a moving object sweeps across the retina, firing a row of photoreceptors in sequence. The eye, too, can lock onto a moving object and follow it. One system detects, the other tracks. But neither explains the stable, balanced world the brain perceives.

On the surface of the brain, in an area called the visual cortex, lie special nerve cells. Linked to rows of the retina's photoreceptors, these cortical cells act as motion detectors. They only respond to an image sweeping across a particular row of photoreceptors. Because the nervous system keeps the brain informed of the body's movements, the cortical cells can distinguish images moving across the retina when the observer moves from those caused by something moving, independent of the eye's shifting vision. While scientists cannot fully explain the eye's extraordinary ability to sift all the movement it encounters, they believe the most important clue in motion detection is the frame of reference, the stationary surroundings of a moving object. A fly crawling on a windowpane is easier to spot when nearer the edge of the pane than the center.

A War of Muscles

The eyes themselves are never truly still. Each eye is equipped with three pairs of taut, elastic muscles. Like reins under constant tension, each of these muscles vies with its counterpart to pull the eye. The superior rectus would roll the eye back, diverting the gaze upward, were it not for the inferior rectus trying to pull the eye down. The lateral rectus would swivel the eye outward were it not for the medial's tugging toward the nose. If one of the oblique muscles were to relax, the eye would roll clockwise or counterclockwise in its socket. The eye is forever caught in the middle of this three-dimensional tug of war.

The constant state of tension helps make this group of muscles one of the quickest in the body. They turn the eyes faster than any neck or other body movement. Capable of seven coordinated eye movements, they give mankind one of the most advanced tracking systems in nature.

A simple illusion demonstrates three of the seven movements — tremor, drift and flick. In a completely darkened room, a small motionless point of light — a lit cigarette on the rim of an ashtray — will appear to hover unsteadily in the darkness. Even if the observer strains to fix his stare on the light, it continues to move. Each goes unnoticed, for all are involuntary. Tremor, an imperceptible trembling, results from the ceaseless warring of the eye's muscles. Drift is responsible for the point of light moving slowly off center.

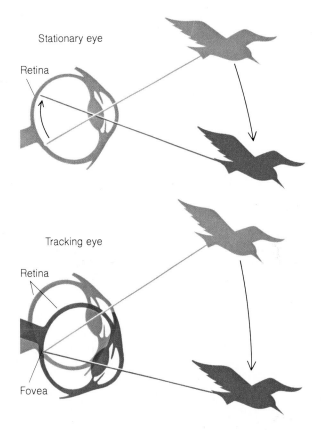

Equipped to detect and track, our eyes can perceive motion in two ways. When the eye remains still, an image sweeps across the retina, firing a row of photoreceptors and signaling the cortical cells on the brain. When the eye locks onto a target and tracks it, the action of the eye muscles informs the brain of motion. The extreme edges of the retina are sensitive only to motion, the most "primitive" form of vision, providing man with a kind of early warning system.

45

Before we are aware of the displacement, a quick flick, the third eye movement, throws the target back on center. These micromovements might seem distracting, but without them we could not see. If an image were held stationary on the retina, the participating photoreceptors would rapidly grow fatigued and the image would fade. Tremor, drift and flick guarantee that an image constantly sweeps across the retina to be processed by fresh photoreceptors.

Smooth pursuit movements mimic drift on a broader scale. Engineered for tracking an object in motion relative to the eye, smooth pursuit requires a real moving target. Ask someone to imagine a flying bird and to follow its path smoothly with his eyes. The person's sweeping gaze is not smooth but jolting; it is a series of small jerks tracing the imagined trajectory. The process of matching the rotation of an inch-wide orb with the velocity of a target "out there" and keeping that target on the fovea is a complicated business. Engineers, building tracking systems for ships at sea, have come close to matching this natural precision that we take for granted. But it is not perfect; our gaze often gets ahead or lags behind the target. To compensate, saccadic eye movement readjusts the eyes' aim.

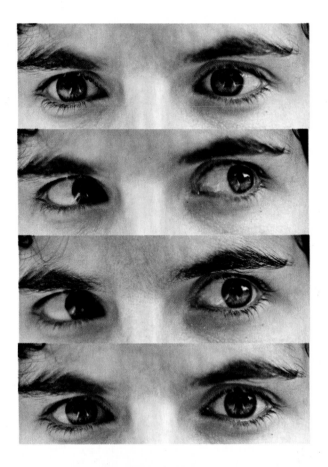

Eye movements can reveal disorders in the nervous system or muscles of the eye itself. At the National Institutes of Health, a patient watches bars of light racing by at controlled speeds. After two-tenths of a second, her eyes involuntarily begin to track the light in an alternating pattern of eye movements, represented on the sawtooth graph. Saccades lock the eyes onto a moving target, while smooth pursuit permits the eyes to drift along with that target. The tracking pattern continues for a few seconds after the lights are shut off — a normal reflex.

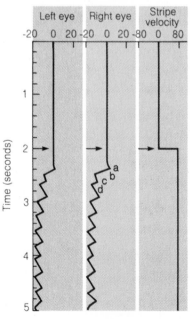

Advertisers use natural eye movements to evaluate the effectiveness of ad campaigns. This study demonstrates the average reader's scanning technique. The pleasant face marred by a black eye grabs the viewer's attention; from there a series of saccades moves the eyes in a typical top to bottom, left to right direction. The importance of the ad's components are placed accordingly.
Toward the end of the scan, the eyes move quickly, spending less than a few tenths of a second on each spot, indicating to advertisers that a point of boredom has been reached and the page is about to be turned.

Saccadic movement darts the eye in the smallest fraction of a second. Programmed by the brain, each saccade requires planning. But the planning seems to take a proportionately long time — a saccade that takes 50 milliseconds to execute can take as much as 200 to plot out. Researchers theorize this delay gives the brain time to send back a signal to the eye canceling the blurred image sweeping by during the saccade.

Saccades are generally voluntary. During reading, they jerk the eyes from one cluster of words to the next, returning the eyes to the beginning of a line in less than 100 milliseconds. When we look at a painting, saccades move the eyes from point to point along the lines of composition. Saccadic movement can also be involuntary. An object detected near the face immediately triggers a saccade, and allows the fovea to pinpoint the potential threat. And, in sleep, while the brain forbids the rest of the body from taking part in the imagined movements of dreams, the eyes participate vicariously in the dream's actions. Saccades known as rapid eye movement, or REM, announce the arrival of dreams.

Smooth pursuit and saccadic eye movements generally work well together, but sometimes their cooperation can be interrupted. Researchers have found that alcohol and barbiturates attack smooth pursuit long before the saccadic, partly explaining why an intoxicated person has difficulty driving. When every object in the visual field is in continuous, gradual motion relative to the driver's eyes, smooth pursuit is essential to giving the brain the information it needs.

That our eyes are a few inches apart means that each retina receives a slightly different picture. When looking at an object about 100 yards away, the eyes are essentially aligned parallel. The image hits the fovea of each eye, creating no noticeable disparity. But as an object approaches, the two images begin to diverge. Were our eyes to remain parallel, we would see everything within 100 yards twice.

This disparity between the two images triggers vergence, the sixth eye movement. Vergence turns the gaze of each eye inward to cast the image on both foveas. A pencil, held at arm's length, will be seen singularly. But moved slowly

toward the nose, as the strain of the eye muscles becomes noticeable, the separate foveal images of the pencil reappear. No further vergence adjustments can be made. The brain monitors the strain on the eye muscles during vergence adjustment and uses this information to estimate distance.

The seventh eye movement is the most complex. A sudden stir on the periphery of vision alerts the eye, which responds with a saccade. If what we see interests us, we turn to face it. The vestibulo-ocular system enables us to keep the object's image on the fovea while the head and body move. One of the body's other senses also contributes to making the complex adjustments. Deep in the ear are three small, crescent-shaped canals, each filled with a viscous fluid and surrounded by nerve cells. These cells form nerve fibers that connect directly to the part of the brain controlling vestibulo-ocular movements. Whenever the head moves, the fluids shift, like bubbles in a carpenter's level. A vertical canal registers up and down movement; fluids of the horizontal canal shift when the head turns right or left; and the tides in the third canal ebb and flow with movement to and fro. By monitoring the shifting fluid, the brain learns of changes in orientation and adjusts the eyes accordingly.

Man's eyes are not superior to other creatures' eyes. They simply suit his needs. The hawk sees at speeds and distances far beyond human vision, while insects see by ultraviolet light invisible to man's naked eye. What makes his eyes unique is his brain, giving him the power to extend the mechanics of his vision. With artificial light, we lengthen the day. With manmade lenses — telescopes and microscopes — we see to the stars and into the living cell. With cameras we extend vision through time and space, witnessing what we were not there to see. By artifice, we enhance our understanding of the world and our place in it.

Chapter 3

Pictures in the Mind

Eyes immobilized and held open with wire clips, head locked in a vise, an anesthetized cat sat passively staring at a screen. As spots of light in changing shapes and sizes flashed on the screen, scientists David Hubel and Torsten Wiesel hovered over the cat. Probing its brain with a microelectrode needle — finer than a hair — they recorded the responses of brain cells to the images on the screen. With their slide show, Hubel and Wiesel broke vision's code in the brain. They discovered some of the basic units in the language the brain uses to interpret the patterns of light striking the eye.

The Visual Pathway

At each step along the pathway from eye to brain, signals carrying visual information link and recombine, fusing their messages into new patterns of information. The signals begin their journey in the retina, a multilayered sheath lining the back of the eyeball. Here lie more than 100 million light-sensitive rods and cones. Each of these photoreceptors detects a minute fragment of the image focused by the lens, then converts the light pattern into an electrochemical signal containing visual information. At one end of each rod and cone is a segment containing light-sensitive pigment; at the other, a synaptic body, or tiny pouch, transmits information by releasing chemical neurotransmitters. Transmitters carry signals to bipolar cells in the retina's middle layer. Bipolars in turn flash impulses to ganglion cells in the retina's innermost layer. Horizontal and amacrine cells relay impulses laterally, integrating visual messages across the retina. Nerve fibers from the ganglion cells form the optic nerve, through which impulses leave the eye.

At the first way station beyond the eye — the optic chiasm — nerve fibers from each eye meet in the brain. From here, each optic nerve splits in half, forming two tracts. Fibers from the inner

Images reaching the eye compete for recognition until the mind can give them meaning. By drawing deductions from a wealth of visual clues, the brain transforms signals from the eye into conscious experience.

51

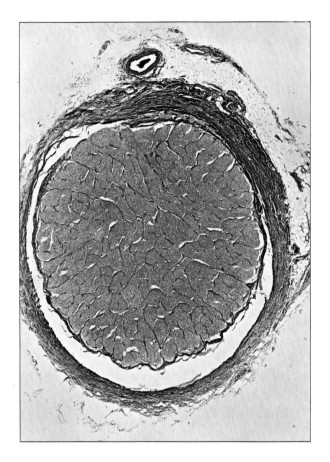

Magnified thirty times, a cross section of the optic nerve resembles a cable, densely packed with more than a million neurons. Formed of the ganglion cells at the back of the retina, each optic nerve divides at the optic chiasm, sending one branch to each of the lateral geniculate bodies in the brain, which, in turn, pass the messages to the visual cortex. It takes less than one-tenth of a second for an image to move along the optic nerve from the retina to the visual cortex, where it is perceived and analyzed.

halves of each retina cross over and head for the opposite hemisphere of the brain. Thus, messages from the right visual fields of both eyes reach the left hemisphere, while signals from the left visual fields of both eyes travel to the right hemisphere. In this way, visual messages from both eyes reach both halves of the visual cortex, the brain's primary sight-processing center. Although each hemisphere receives images from only one half of the visual world, the brain merges them into an integrated whole. The partial crossing of nerve fibers at the optic chiasm accounts in part for human stereoscopic, or three-dimensional, sight. In lower animals like the frog, the optic nerves cross completely. All nerve fibers from each eye travel to the opposite hemisphere, thereby making three-dimensional vision impossible.

The next stops on the visual pathway are the lateral geniculate nuclei, twin relay stations deep in each brain hemisphere. Some scientists think the lateral geniculate nuclei coordinate visual signals with other sensory information, since other sense organs also send impulses to them. According to psychologist John Frisby, the lateral geniculate bodies may be "involved in shutting out visual inputs to the brain when attention is being devoted to some other source of information."

Most visual impulses travel directly from the lateral geniculate nuclei to the primary visual cortex, also called the striate cortex because its many layers give it a striped appearance. Our ability to detect the spatial organization of a scene and the shapes of objects — the brightness and shading of their parts — depends upon the functioning of the striate cortex. From here, some nerve fibers project to surrounding areas known as the prestriate, or secondary visual, cortex. Scientists believe these regions decode visual messages at a higher level than the striate's. The prestriate may contribute to pattern recognition — the ability to identify a flower as a flower because damage to this brain region often results in the inability to perceive patterns.

From the prestriate cortex, visual impulses enter the temporal lobes located near the sides of the head for further — and probably more sophisticated — processing. Damage to visual sites in the temporal lobe can inhibit visual learning.

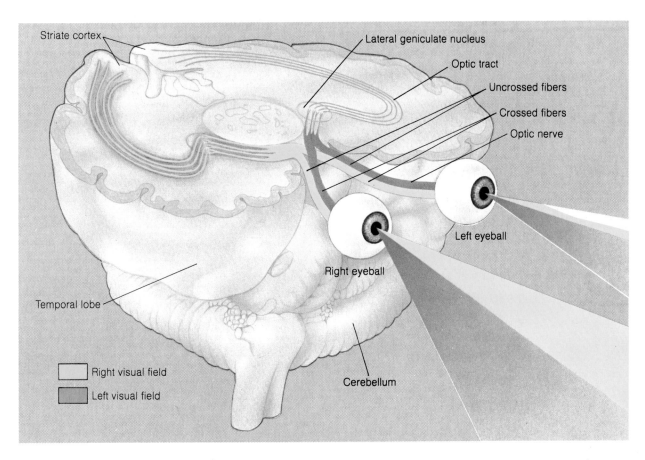

Striate cortex

Lateral geniculate nucleus

Optic tract

Uncrossed fibers

Crossed fibers

Optic nerve

Left eyeball

Right eyeball

Temporal lobe

Cerebellum

Right visual field

Left visual field

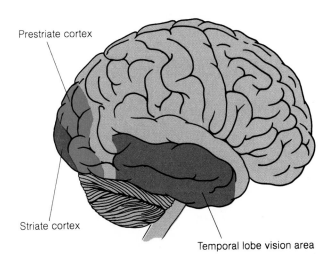

Prestriate cortex

Striate cortex

Temporal lobe vision area

As light strikes the photoreceptor cells in the retina, electrochemical impulses begin their journey along the visual pathway. These messages are carried by the optic nerves, which divide at the optic chiasm, sending signals from each eye to both hemispheres of the brain. After passing through the lateral geniculate bodies, most visual messages travel first to the striate cortex, shown here in red. Scientists believe the striate cortex organizes objects spatially and gives them shape and brightness. Some impulses then travel to the prestriate cortex, in blue, which may help us recognize patterns. It is believed the temporal lobes, shown in green, contribute to visual recognition and memory.

Computer-constructed images of a monkey's visual cortex tell scientists how the brain interprets visual information. Glowing with radioactive glucose, yellow and red areas indicate high visual activity.

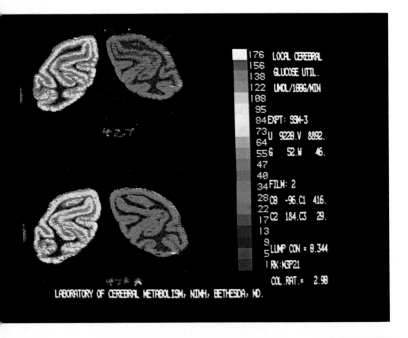

LABORATORY OF CEREBRAL METABOLISM, NIMH, BETHESDA, MD.

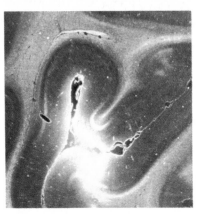

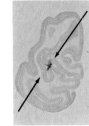

Winding their way through the brain as if carried on a river, amino acids follow neural routes in the visual cortex. Building blocks for protein, the acids travel from cell to cell. A cross section of the brain of a rhesus monkey reveals bright spots where scientists have injected radioactive amino acids into the striate cortex, the primary sight-processing center. The arrows in the diagram indicate the points to which the amino acids have flowed.

When Chicago psychologists Heinrich Klüver and Paul Bucy removed a monkey's temporal lobes, the animal lost its ability to recognize objects by sight. It also lost its fear of approaching objects it had previously avoided. This phenomenon, called psychic blindness, or the Klüver-Bucy syndrome, sometimes occurs in humans who have suffered damage to the temporal lobes.

New techniques for mapping visual circuitry show that visual impulses travel to realms of the brain involved in behavior and emotions. At the National Institutes of Health, researchers have traced the visual pathway into a midbrain region known as the limbic system. Pioneered by Louis Sokoloff, director of the Laboratory of Cerebral Metabolism at the National Institute of Mental Health (NIMH), these methods map brain sites with a form of glucose, the substance brain cells use for energy. Mortimer Mishkin and colleagues at NIMH's Laboratory of Neuropsychology "blind" monkeys on one side of the brain by severing nerves relaying signals from the retina. The scientists then cut connections between the brain's left and right hemispheres, arresting the flow of information between the two halves. After being injected with radioactive glucose, the monkeys watch revolving patterns to stimulate the visual areas of their brains. The glucose diffuses throughout the brain, settling in greater concentrations in more active regions. X-ray films of slices of the monkeys' brains reveal the pattern of glucose use. On a computerized color map, says Mishkin, "the differences . . . stand out in beautiful color." The research team found high concentrations of glucose in the amygdala, a structure of the limbic system that plays a role in fear and aggression. According to Mishkin, these connections might explain why "when we see things that are fearful, we are afraid."

Visual fibers also travel to another midbrain center, the superior colliculus, which seems to guide visual attention. When an object appears on the periphery of our vision, the circuitry of the superior colliculus detects its presence and signals the eyes and head to turn so that we can focus on the object and identify it.

The job of examining and identifying an object belongs to the visual machinery of the cerebral

David Hubel and Torsten Wiesel

Cracking Vision's Code

Describing a kitten chasing its tail, Ralph Waldo Emerson wrote, "If you could look with her eyes you might see her surrounded with hundreds of figures performing complex dramas, with tragic and comic issues, long conversations, many characters." Man and cat, no doubt, see very different worlds. But as two Harvard researchers have shown, the process of vision in cats offers clues about the nature of vision in man.

Neurobiologists David Hubel and Torsten Wiesel began their research in the late 1950s. They placed a cat before a screen, eyes open, and projected shapes of light onto the screen. Using microelectrodes, they recorded the responses of individual neurons to spots and bars of light. The ganglion cells underlying the retina's photoreceptors possessed specific receptive fields on the surface of the retina that regulated each cell's response to light. The ganglion's response was strongest when a small spot of light struck the center of its receptive field.

Hubel and Wiesel continued to probe along the pathway of sight, encountering neurons in the brain that fired more or less frequently depending on the point at which a small spot of light touched the retina. But

when the scientists probed the striate cortex, the brain's primary visual processing center, they discovered a remarkable difference in the behavior of the cells. There, neurons responded to specific shapes and angles of light — usually a bar of light or an edge created by adjoining light and dark areas. A bar of light projected at an angle corresponding to two o'clock on the face of a watch

might trigger a cell to fire. But if the angle changed to five o'clock, the cell would stop firing and another cell would begin. Hubel and Wiesel called these neurons "simple" cells.

Probing deeper into the striate cortex, they discovered another group of neurons which they named "complex" cells. These cells had broader receptive fields. They responded to a properly slanted bar of light located anywhere in their receptive field. Some complex cells continued to respond if the stimulus moved in a specific direction. Others reacted only to short bars of light or to bars that met at a ninety-degree angle to form a corner. "The number [of neurons] responding successively as the eye watches a slowly rotating propeller is scarcely imaginable," Hubel marveled.

Although their work won them a Nobel Prize in medicine in 1981, Hubel and Wiesel warn that probing the brain's visual processing mechanisms falls far short of explaining vision. "Our conclusions by no means solve the main question in vision — that is, how the brain makes sense of the image that falls on the retina, the image of a scene rich in form, color, depth and movement. We are only beginning to see a few steps along the way."

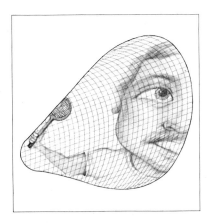

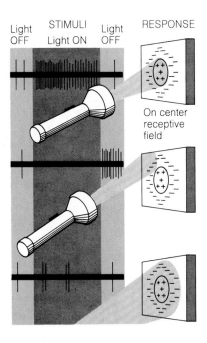

STIMULI

Light OFF   Light ON   Light OFF

RESPONSE

On center receptive field

By probing individual neurons with fiber-thin electrodes, scientists have discovered how ganglion cells in the retina respond to light that strikes their receptive field — an area of photoreceptors on the retina's surface. Neurons in the visual cortex of the brain begin to fire when light strikes the sensitive center of an "on-center" cell's receptive field, represented above. Light outside the center of the receptive field inhibits the neuron's response.

cortex, the brain region responsible for man's higher functions. Visual messages first reach the striate cortex, located in the occipital lobe at the back of the head. By charting the responses of visual cells to bars of light slanted at various angles, Hubel and Wiesel identified the triggering stimuli for hundreds of brain cells. Each neuron, or nerve cell, in the striate cortex has its own receptive field — an area on the retina that excites or inhibits the firing of that neuron.

Their experiments with the striate cortex led Hubel and Wiesel to formulate an important theory — that visual information is processed in a hierarchical fashion. Because each nerve cell connects with many others at each stage of the visual pathway, the messages undergo constant transformation, and the complexity of the information increases at each level of the visual circuitry.

Feature Detectors

At the beginning of the visual pathway, cells respond to simple features. Hubel and Wiesel discovered that retinal ganglions respond best to disks of light — either a dark disk surrounded by a light area or a light disk on a dark ground. Cells in the lateral geniculate bodies also respond to spots of light against contrasting backgrounds. Retinal ganglion and geniculate cells have doughnut-shaped receptive fields. Some send impulses when light strikes the center of their fields. They slow down or stop firing when the surrounding area is lit. These are "on center-off surround" cells. In other cells, the opposite occurs. Light in the center of their receptive fields inhibits them, while illumination of the surrounding regions excites them to fire. Such neurons are "off center-on surround" cells.

In both the striate and prestriate areas of the cortex, the visual world is mapped out point for point, each tiny cortical segment corresponding to a specific section of the retina. Because of their importance, however, images focused onto the foveal, or central, region of the retina have far more detailed representation in the cortex than do images located on the periphery of the visual field. Thirty-five times as much space is devoted to the fovea, giving it a greater degree of visual acuity than any other region of the retina.

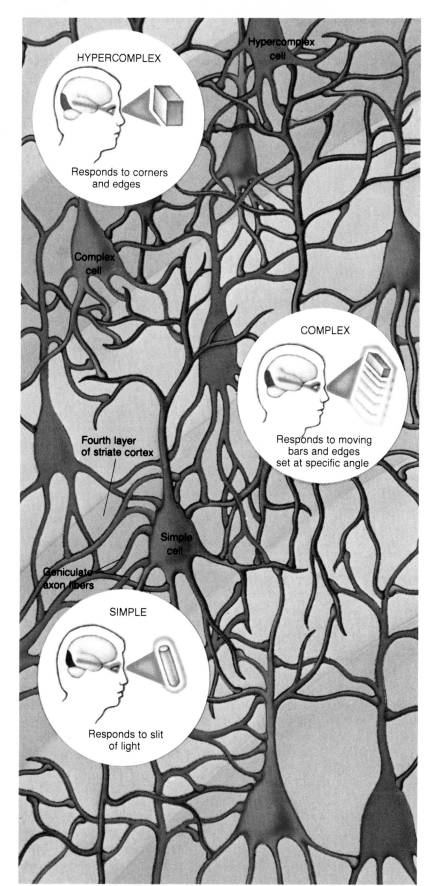

HYPERCOMPLEX

Responds to corners
and edges

Hypercomplex
cell

Complex
cell

Fourth layer
of striate cortex

COMPLEX

Responds to moving
bars and edges
set at specific angle

Simple
cell

Geniculate
axon fibers

SIMPLE

Responds to slit
of light

Like rows of fibrous roots, axons and dendrites of visual cells in the brain catch electrochemical impulses relayed through nerve connections called synapses. According to one scientific theory, visual cells are hierarchically arranged by function. Simple cells recognize precisely positioned slits of light. Complex cells detect slanted lines and movement in one direction. Hypercomplex cells identify corners and angles. In the striate cortex, cells activated by stimuli tilted in the same direction form vertical columns. The visual cells of frogs, humorously called "bug detectors," respond best to small objects moving rapidly. The cells are so specialized that a frog can starve in a field of dead flies.

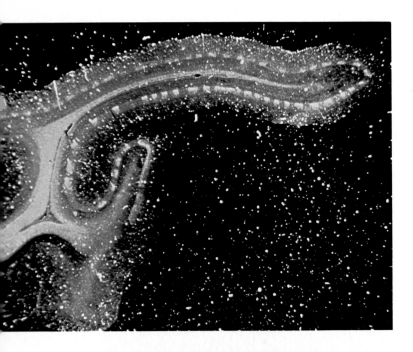

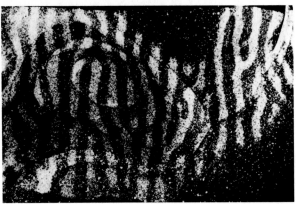

By injecting radioactive nutrients into the fluid of the eye, scientists can trace the architecture of the striate cortex. In these autoradiographs of the visual cortex of a macaque monkey, seen from above, top, and from the side, below, radioactive amino acids illuminate the ocular dominance columns, groups of cells influenced primarily by one eye. The bright columns show right-eye preference; the dark columns respond more to the left eye.

Like retinal and geniculate cells, visual cells in the cortex also detect certain features of light, but their standards are more exacting. Some respond only when three or four features are present. Scientists divide cells of the striate cortex into three main categories. "Simple" cells respond best to a slit of light on a contrasting ground or to boundaries (edges) between dark and light zones. The slit or edge must be precisely located in the cell's receptive field and must have the proper orientation — positioned vertically, horizontally or at a precise angle. The receptive fields of simple cells are also divided into "on" and "off" areas, but they are bar-shaped rather than circular.

At the next level of the visual hierarchy are the "complex" cells. They also respond to slits and edges set at specific angles. Unlike a simple cell, however, the complex cell will fire in response to a slit located anywhere in its receptive field or to a bar moving in a specific direction. This property represents a higher level of visual processing: the ability to distinguish shapes regardless of their location. "Hypercomplex" cells respond best to features meeting an even more refined set of requirements: corners, angles, lines or edges of a specific length, orientation and location. Some also respond to movement in one direction.

The hierarchical network of cell connections in which complicated cell properties build up from much simpler cells suggests an extremely orderly cellular architecture. In order for a complex cell to receive input from a group of simple cells sharing the same orientation requirements, the two must be located near each other. This is exactly what Hubel and Wiesel discovered. Pushing an electrode through the striate area in a direction perpendicular to the surface of the cortex, they found that every cell within a certain depth was stimulated by bars tilted at the same angle. A thousandth of an inch to either side, cells responded to a different orientation. Hubel and Wiesel named these units orientation columns. They found similar columns connected to each eye. Shaped like vertical slabs, these ocular dominance columns consist of thousands of cells receiving information primarily from the same eye.

From the visual cortex, a special circuit of nerves carries signals to the cerebellum. Lying

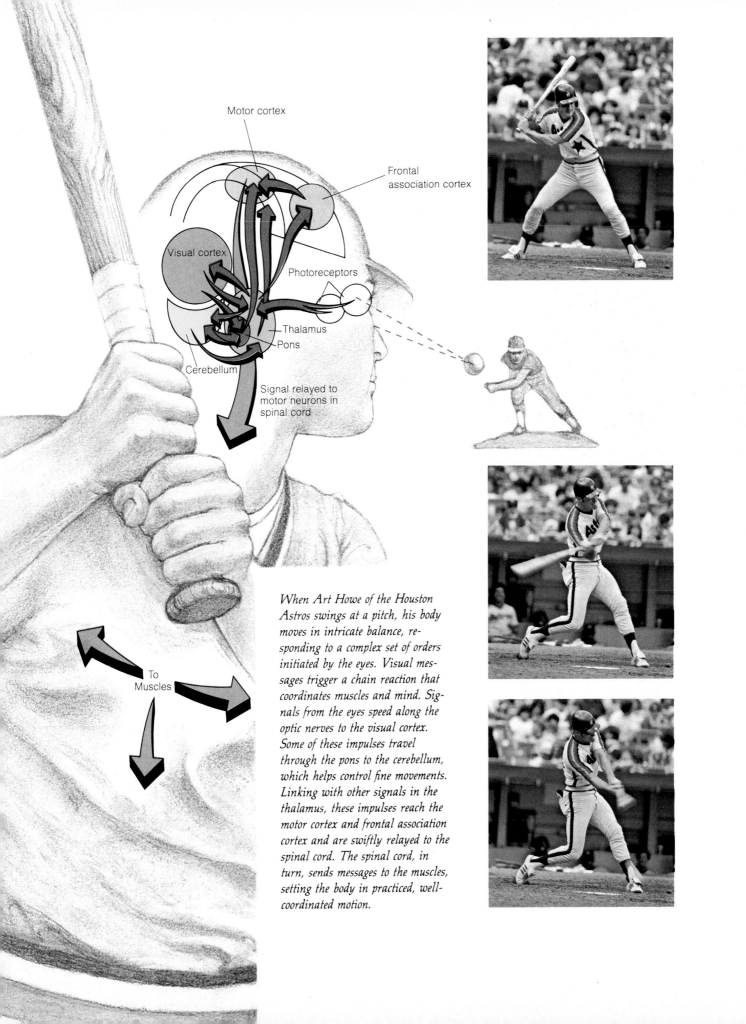

Motor cortex

Frontal
association cortex

Visual cortex

Photoreceptors

Thalamus

Pons

Cerebellum

Signal relayed to
motor neurons in
spinal cord

To
Muscles

When Art Howe of the Houston Astros swings at a pitch, his body moves in intricate balance, responding to a complex set of orders initiated by the eyes. Visual messages trigger a chain reaction that coordinates muscles and mind. Signals from the eyes speed along the optic nerves to the visual cortex. Some of these impulses travel through the pons to the cerebellum, which helps control fine movements. Linking with other signals in the thalamus, these impulses reach the motor cortex and frontal association cortex and are swiftly relayed to the spinal cord. The spinal cord, in turn, sends messages to the muscles, setting the body in practiced, well-coordinated motion.

A kitten peers into a restricted world, seeing horizontal stripes with its left eye, vertical stripes with its right eye. Helmut Hirsch and his colleagues have learned from these experiments that brain cells can be so altered by early visual experience that kittens have difficulty walking on surfaces their restrained vision cannot recognize. Other experiments have shown that the kittens' altered visual cells can partially adapt to an unrestricted environment.

below the occipital lobes, the cerebellum coordinates muscle movements. This visual pathway regulates activities requiring close coordination between the eyes and limbs, such as driving a car or catching a ball. By blending motor and visual signals, the cerebellum controls the position of the body and limbs as a person moves. Scientists have discovered that some of the visual impulses make a detour on their way to the cerebellum. They travel through the pons, a bridgelike structure at the base of the brain. Certain cells in the pons respond to single spots of light moving at specific rates and directions. Other cells register large, moving visual fields containing many spots of light. These cells send new signals to the cerebellum, which synthesizes them with information about the position and velocity of muscles and joints.

Altering the Circuitry

How much of this complex visual circuitry is present at birth? Can it be influenced by abnormal visual experiences during infancy or childhood? In 1970, British physiologists Colin Blakemore and Grahame Cooper sought an answer to the "nature-nurture" issue. They raised two kittens in separate chambers. One chamber was lined with vertical stripes, the other with horizontal stripes. The stripes became the kittens' sole form of visual stimulation for five months. Afterwards, the kittens were moved into a laboratory with horizontal and vertical surfaces. The kitten that had been "horizontally deprived" while living in the vertically striped chamber could not see horizontal surfaces. The other kitten, accustomed to horizontal lines, easily jumped from one horizontal surface to another, but walked into chair legs and other vertical obstacles. The scientists found that most cells in the visual areas of the kittens' brains responded only to light of the orientation that had dominated the kittens' environment. The first kitten lacked nerve cells for detecting horizontal lines; the second lacked vertical line-detecting cells.

At about the same time, Helmut Hirsch and D. N. Spinelli of Stanford University conducted a slightly different version of this experiment. To ensure that head movements would not alter the images, the scientists outfitted the kittens with goggles. One lens contained a picture of vertical stripes; the other, horizontal stripes. After twelve weeks, Hirsch and Spinelli discovered that visual cells in the animals' brains had been altered by the abnormal stimulation. Most cells responded only to horizontal or vertical stimuli instead of the full range of angles that stimulate visual cells in animals raised in a normal environment.

Hubel and Wiesel then sutured a newborn monkey's eyelid for several weeks. They found that the animal remained nearly blind in that eye after the removal of the stitches. The ocular dominance columns corresponding to the closed eye had shrunk, while columns responding to the sighted eye had expanded. Like such monkeys, human infants born with amblyopia —impairment of vision in one eye — retain permanent impairment in the weak eye if not treated by ages six to ten. By then, some scientists think, shrinkage of the ocular dominance columns corresponding to the weak eye cannot be reversed.

The visual system is most vulnerable to environmental influences during an initial visual learning period that ranges from three months in cats to five or six years in humans. Scientists at the California Institute of Technology have discovered a brain chemical that appears crucial to learning during this period. They removed this substance, a catecholamine, from the brains of kittens less than six weeks old, then sewed together one eye in each kitten. Normally, the blocked eye would be permanently impaired. But when the scientists removed the stitches, the kittens retained normal vision. The lack of catecholamine had apparently rendered the kittens incapable of visual learning. The research team also discovered that an injection of catecholamine increases the brain's plasticity and may promote visual learning in animals far beyond the learning stage. Eventually, catecholamine may be used to cure people with impaired depth perception.

Building a Visual World

While neurophysiologists explore the brain's visual terrain cell by cell, other investigators seek to discover how the mind transforms the dots

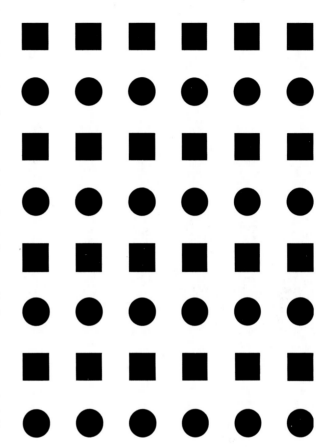

Unthinkingly, we arrange the squares and dots horizontally in their own groups, although they also form vertical rows of mixed components. Gestalt, the German word for an array of shapes perceived as a whole, has come to signify an entire branch of psychology. Important as a principle of visual organization, the Gestalt theory of similarity shows that the mind attempts to order random patterns by linking related items in a group.

and lines the eye sees into a visual impression of the world around us. Why do we see a moonlit beach or a smiling child instead of a hodgepodge of light patterns? Hubel, Wiesel and other scientists have identified the building blocks of perception — the feature-detecting cells of the retina, lateral geniculate nuclei and visual cortex. But researchers do not yet know how these blocks combine to form perception of scenes.

One theory, advanced by the Gestalt school of psychology in the early part of this century, suggested that certain principles of perceptual organization dictate the way we see the various parts of an image. Imagine several rows of dots, spaced more closely in a horizontal direction than vertically. The horizontal rows will probably be perceived as parts of the same group. This illustrates the Gestalt law of proximity: elements close to one another are more likely to be grouped by the eye than those farther apart. This law embraces other senses as well. A young child, unaware of the relationship between thunder and lightning, nonetheless perceives them as parts of the same event because the loud clap and the flash of light occur almost simultaneously.

According to the Gestalt law of similarity, we perceive similar elements as belonging together.

Two large circles will stand out in a field of small dots. A third law, grouping by "good form," states that elements forming a "good figure" will be perceived together. A good figure is symmetrical, completed and composed of straight or curved lines.

The ability to recognize patterns is crucial for scanning and reading. When searching a page for a specific word — such as a name in a telephone book — we do not identify, or even notice, each word. We look for distinctive features of the target word, such as the first letter. Many scanning operations can take place at the same time. When Cornell University researcher Ulric Neisser tested scanning skills, he found that readers were able to detect four target letters on the same page as quickly as one. After a two-week training period, subjects could carry out a ten-target letter scan just as speedily. This is possible, Neisser proposes, because word and letter searches take place at a relatively low cognitive level, using only rudimentary visual processes.

Similarly, we need not read every letter or word in a sentence to perceive its meaning. A skilled reader can "see" only three or four letters a second. If he had to read every letter of every word to understand written material, his reading

Although only one of these Coca-Cola bottles has English writing on it, we recognize them instantly. Our visual memory and cultural experience make the world familiar to us, giving objects their proper context.

Many psychologists think certain visual skills are inborn since some blind people can depict shapes, perspective and motion. Blind from birth, Jane Welliver carved this striking figure from paraffin.

speed would be 35 words a minute — far below the average speed of 300 words a minute. Instead, the reader perceives the meaning of sentences by recognizing certain patterns in the text — features such as word length and grammatical parts of speech such as nouns, pronouns and verbs. Combining these features, the reader formulates an internal grammatical message. This process takes place unconsciously, mirroring what goes on in a person's mind as he speaks or tells a story.

In some cases, a message can be conveyed solely through a visual image. The general context in which an image appears can transform a single symbol into a message that can lead the mind to make other associations. Highway signs for public telephones, motels and restaurants rely on simple representational diagrams to remind drivers that roadside accommodations are available. For the 1968 Summer Olympic Games in Mexico City, graphic artists devised signs using a shower nozzle, coat hanger, standard envelope and telephone dial to represent various public services. Two images of a pipe with smoke wafting from its bowl — but one with a diagonal bar drawn through it — indicated smoking and nonsmoking sections of buildings and sporting arenas.

These symbols are all readily understood by Americans, Europeans and other people familiar with Western culture. A tribesman from Africa or the Pacific Islands, however, would only be puzzled by the designs. Because his world has no showers or telephones, he would not comprehend the symbols. Art historian E. H. Gombrich has explained that our understanding of symbols and other visual images "always depends on our prior knowledge of possibilities." Language and the culture it derives from work together with visual impressions to reinforce memory and give meaning to messages the brain receives. It is through images striking the eye that the mind gives shape to its surroundings.

Studies on the recognition of faces reveal that most people can easily detect a familiar face in a crowd. On the other hand, we often make poor eyewitnesses. The perceptual skills used in recognizing a familiar face — detecting and assessing certain features — are often ineffective when used to identify someone seen fleetingly. Percep-

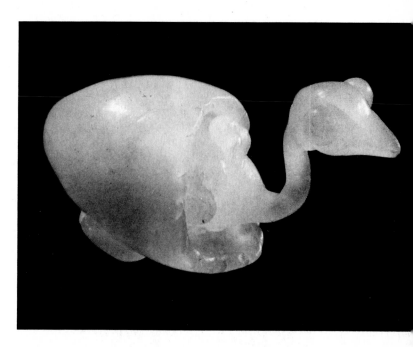

tion, like memory, does not work like an instant replay machine. Both are selective, decision-making processes affected by a person's background, attitudes, skills and environment.

Nature or Nurture

Throughout history, the nature-nurture issue has played a prominent role in the study of perception. Empiricist philosophers John Locke and George Berkeley maintained that perceptual skills were not innate, but were acquired through learning and experience. Nativists such as René Descartes took an opposing view, claiming that humans were endowed with perceptual abilities at birth. Even after centuries of debate and experiment, scientists still seek ways to measure the contribution of both influences. Researchers today pursue several avenues in an attempt to sort out what is programmed into the genes from the influences of the environment. One avenue of study involves the blind and the formerly blind.

After nearly a decade of working with blind patients, John Kennedy, a psychologist at the University of Toronto, has concluded that the blind have an intuitive sense of perspective and are capable of recognizing shapes and understanding visual ambiguities. Kennedy's findings

suggest that some perceptual abilities are innate. His patients were able to describe and draw objects. Their drawings ranged from depictions of a hand to moving objects and a cube drawn in perspective. Their portrayal of moving objects was especially remarkable. Many used modern artistic devices to illustrate motion, including bent limbs and a line of movement at the feet of a running figure. Few of the patients had had any previous drawing experience. They had developed these artistic solutions on their own.

That humans are genetically endowed with some degree of visual perceptual skills has been demonstrated by the experiences of people blinded from birth or early childhood by cataracts or clouded corneas, and later surgically cured. At first bewildered by the sudden bombardment of visual imagery, these individuals were nonetheless soon able to perceive figure-ground relationships, to scan and focus on objects and to track moving objects with their eyes.

Yet, most of the formerly blind remain highly dependent on their sense of touch. The case of "S. B." illustrates the difficulties some people have in adjusting to a visual world. His sight surgically restored at the age of fifty-two, S. B. still relied largely on his sense of touch. Although he eventually grew to recognize certain objects by sight, S. B. could not retain visual information or incorporate it into his drawings. Less than two months after his operation, he drew a picture of a bus, but was frustrated by his inability to draw the radiator — one of the few parts he had not touched when blind. His depth perception was poor at first. He believed his feet would touch the ground outside his hospital window if he sat on the sill, although the ground was thirty feet below. He was terrified to cross the street — a task he had performed fearlessly while blind. With his newfound sight, S. B. realized how blindness had robbed him of many opportunities and experiences. Saddened, he withdrew from the sighted world and lived the rest of his life almost as a blind man again, often not bothering to turn on the lights in the evening.

Studies of the blind, however, do not provide the ideal means for determining the relative influences of heredity and environment on the visual process. Blind people have accumulated a vast store of knowledge from other senses to help them perceive the world. Perhaps more revealing are perceptual studies of infants. Babies only a week old display visual skills. Even at such an early age, their eye movements are not random. They prefer features that help define an object, such as edges and contours, over simpler shapes. Infants also seem to prefer the human face. Given a choice between an abstract drawing or a picture of a face, a two-month old baby spends about twice as long looking at the face.

In the "visual cliff" experiment, babies between six and fourteen months old showed the ability to perceive depth. Placed on a central bridge between a normal floor and a clear glass surface suspended several feet above a patterned floor, the babies refused to crawl onto the glass side. Even their smiling mothers could not entice them over. Chickens, goats and lambs only one day old also chose to remain on the normal side.

How do visual images striking the flat, two-dimensional surface of the retina become transformed into the three-dimensional world we ultimately perceive? Somehow, the brain synthesizes information about depth and distance from a number of sources, weighs them against each other and arrives at an impression of how objects are located in space. In some cases, we use information from just one eye to judge depth, as when viewing a picture or observing a scene from more than 200 feet away. One monocular depth cue is called superposition, the overlapping of two or more objects. An object that obscures the view of another is perceived as being closer. Differences in texture serve as another monocular depth cue. The fewer details an object has, the further its perceived distance. The height of various objects lying along a flat plane indicates their relative distance, the furthest object appearing highest.

Accommodation is a monocular depth cue derived from the mechanics of the eye. When we attempt to focus on an image, the lens of the eye must adjust to the proper curvature — it becomes flatter for distant objects, more curved for nearby objects. Sense organs in the muscles that control the curvature of the lens provide the brain with information about its shape. The brain translates

The mind reaches out for visual cues in order to assign correct placement and proportion to scenes that the eye takes in. The pathway and ceiling vaults of this colonnade in Dharan, Saudi Arabia, converge at a visual vanishing point behind the man. If we were to cover the portion of the photograph above the man's head, however, we would be unsure whether he were walking along a flat path or ascending a staircase. The vaulted archway, receding from the pebbled columns close in view, gives the picture its correct — and dramatic — perspective. Similarly, we form a mental image of the entire beach at Valhalla, British Columbia, although we can accurately judge the size and texture only of the stones nearest the camera, opposite bottom. Another depth cue, superposition, occurs when figures overlap one another. Thus, we see that the bowl and scales are at the front of the shop room and that, of the two totem poles, the winged one stands closer to the viewer, opposite, top.

Invented by British physicist Charles Wheatstone in 1838, the stereoscope used the newly discovered principle of stereoscopic vision. We see a three-dimensional image because our eyes observe a scene from slightly different angles. In this stereoscope picture from 1898, left, the corner of the room is in a different position behind the figures in each view; the milk bottle behind the policeman can barely be seen in the left-hand photograph.

With their visual fields overlapping at slightly different angles, the eyes form complementary images of the same object. The visual cortex merges the two views into a single three-dimensional image.

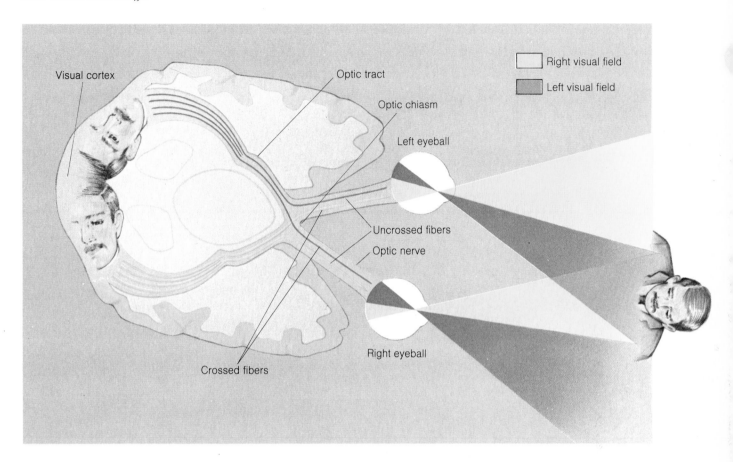

these messages into information about distance. Accommodation is useful only up to twenty feet. The amount of lens adjustment occurring beyond that distance is minimal.

But monocular vision does not provide the full experience of depth. For this, we need two eyes. The fusion of two images viewed from slightly different angles produces a three-dimensional, stereoscopic view. Known also as retinal disparity, stereoscopic vision is the most powerful binocular depth cue. When we view objects within a distance of 200 feet, the binocular depth cue known as convergence comes into play. As we attempt to focus on an object, the eyes swing inward until their lines of sight are directed toward the image. The angle of this swing provides a subconscious clue to the distance of the image. The closer the image, the greater the angle. At distances greater than 200 feet, convergence is ineffective because the lines of sight for both eyes are nearly parallel. For very close objects, convergence also loses its usefulness. At close range, the eyes cannot merge their visual fields.

A cue called motion parallax helps us determine the distance of moving objects. Nearby objects pass the viewer more rapidly than distant ones. By comparing speeds, the mind can judge which is closer. When gazing out the side window of a moving car, a church steeple a mile away will remain in view far longer than a tree by the side of the road.

These principles of stereoscopic vision were not discovered until the nineteenth century. In 1838, British physicist Charles Wheatstone concluded that both eyes could not receive an identical view of the same object. He built an instrument called a stereoscope, which uses mirrors to present different drawings to each eye. The left eye receives a left eye's view of a scene; the right eye sees the same scene but viewed as

Like a large abstract sculpture, a
shell rests on the floor of what
appears to be a beach cottage. Visual
cues such as the walls meeting in
corners, the open door and the sea-
scape tell us that we are looking
across a room, but the shell is
actually normal-sized and has been
placed in a doll house. The perceptual principle of size constancy
allows us to see objects in their
proper size by relating them to their
accustomed surroundings.

only the right eye can see it. Combined, the drawings take on a depth that is missing when they are viewed with only one eye.

Seeing Is Not Believing

In the perception of size, seeing is not really believing. The mind must somehow distinguish an object's true size from its size on the retina. Despite the fact that the image on the retina grows or diminishes according to distance, we think of an object as remaining the same size no matter how far away it is. The perception of size demonstrates constancy — an important feature of the perceptual experience. A given book is perceived as being ten inches long whether held or on a bookshelf across the room. If we were to use depth cues to judge its size from across the room, however, the book would seem much smaller than ten inches. With unfamiliar objects, depth cues often aid in making size judgments. The more familiar an object, the more we judge these features from our experience. The constancy principle also applies to other visual features, including color, shape and location, all of which appear to change under certain conditions.

The ability to separate a figure from its background is perhaps the most basic type of perceptual organizing, exhibited by both newborn babies and blind people whose sight has been restored. The mind almost always perceives one part of the visual field standing out from the rest. Areas of an image that are simpler, more regular, better defined or more solid are more likely to be perceived as figure than ground.

The influence of culture and environment on visual perception was explored by Robert Laws, a Scottish missionary working in Malawi, Africa, during the late 1800s. He wrote:

> Take a picture in black and white and the natives cannot see it. You may tell the natives, "this is a picture of an ox and a dog," and the people will look at it and look at you and that look says that they consider you a liar. Perhaps you say again, "Yes, this is a picture of an ox and a dog." Well, perhaps they will tell you what they think this time. If there are a few boys about, you say: "This is really a picture of an ox

The eye deceives the mind when an object and its background vie for recognition. In an ingenious illustration of the perceptual principle known as figure and ground, the profiles of Britain's Prince Philip and Queen Elizabeth greet each other from opposite sides of this porcelain vase created for the Queen's Silver Jubilee in 1977. The royal faces reveal themselves as figures, transforming the vase into a white background.

What is above the woman's head? When scientists showed a similar sketch to people from East Africa, nearly all of the participants in the experiment said she was balancing a box or metal can on her head. In a culture containing few angular visual cues, the family is seen sitting under a tree. Westerners, on the other hand, are accustomed to the corners and boxlike shapes of architecture. They are more likely to place the family indoors and to interpret the rectangle above the woman's head as a window through which shrubbery can be seen.

and a dog. Look at the horn of the ox, and there is his tail!" And the boys will say: "Oh! yes and there is the dog's nose and eyes and ears!" Then the old people will look again and clap their hands and say, "Oh! yes, it is a dog."

Only the most isolated populations seem totally incapable of comprehending pictorial material. But most evidence suggests that people who have never seen a picture or a photograph must "learn" how to see the objects in them.

The ability to recognize objects in pictures is not the only perceptual skill that varies in different cultures. The perception of depth in drawings eludes many non-Westerners. Pictures containing intersecting lines in which a Westerner would see depth appear flat to people living in environments with few rectangular objects.

Western man's highly "carpentered" world, abundant in boxlike buildings and other rectangular shapes, may shape his visual biases as well. Canadian researchers Barrie Frost and Robert Annis discovered this while comparing the visual skills of city dwellers with those of Cree Indians living in teepees on the eastern shore of James Bay in Quebec, Canada. The Cree's environment has few rectangular objects. Tested with a device

George Stratton

Standing the World on Its Head

During his moonlit walks, George Malcolm Stratton must have presented a bizarre sight to his neighbors as he strode down the path with a plaster cast on his head. Strange as it appeared, this attire enabled Stratton to investigate the visual process by standing the world on its head.

Light entering the eye produces an inverted image on the retina. But Stratton, a psychologist, and other scientists of the late nineteenth century wondered if the inverted image was vital to seeing the world right-side up. To find an answer, Stratton fashioned a plaster cast that fit snugly over his eyes and held a telescope-like tube in place. Four lenses in the tube reversed images from top to bottom and right to left, resulting in an upright image on the retina. For eight days, he wore this odd contraption whenever he was awake and replaced it with a blindfold when he slept.

The first day Stratton wore the lenses he was confused. "The entire scene appeared upside-down," he wrote. "Almost all movements performed under the direct guidance of sight were laborious and embarrassed." By the second day, although exhausted and nauseated at times, Stratton found he could perform simple

tasks, such as washing his hands, with greater ease. He slowly became more adept at picking up an object with the correct hand and returning it to its proper place. Although he grew more comfortable in the world of objects, however, Stratton's perceptions of his body remained complicated. Without looking at his arms and legs, he would sometimes visualize them in "the newer sight" of his inverted world. At other times, the old image of his body would completely override the new.

During the fifth and sixth days of his test, Stratton noticed that actively involving himself in his surroundings seemed to make them more harmonious. By the seventh day, Stratton was "more at home in the scene than ever before. There was perfect reality in my visual surroundings, and I gave myself up to them without reserve. . . ."

By the eighth day, even his sense of hearing was better attuned to the lenses. If he saw a tapping pencil at the left, the sound seemed to come from the left as well. Stratton's perceptions of his body continued to shape his impressions of his surroundings. "As long as the new localization of my body was vivid, the general experience was harmonious," the psychologist reported. But if the old positioning of his body came to mind, the harmony vanished and Stratton "seemed to be viewing the scene from an inverted body."

Shortly after noon on the eighth day, he removed the lenses, ending the bizarre experiment. "On opening my eyes, the scene had a strange familiarity," he wrote. Within hours Stratton had nearly regained his original perspective. He experienced only a few twinges of strangeness over the next few days, reminders of his senses' closely knit effort to find stability in a world turned upside-down.

that displayed lines of every angle, the Crees showed no preference for any particular type of line. Urban test subjects, however, always preferred vertical or horizontal lines.

Feedback for Flexibility

The adult visual system is remarkably flexible. It can adapt quickly to a wide variety of unusual conditions. In experiments to test visual adaptability, subjects have been able to lead normal lives while wearing lenses with prisms, mirrors and other optical devices that turned their perceptual universe topsy-turvy. In 1896, University of California psychologist George Stratton wore goggles equipped with prisms that inverted the retinal image. The prisms turned his world upside down and reversed it from left to right. After the first few dizzying days, in which even the simplest task posed an almost insurmountable challenge, he began to perceive the world as sometimes being right-side up. "When I looked at my legs and arms," he reported, "what I saw seemed rather upright than inverted."

Stratton's peculiar experience has since been duplicated many times. Austrian researcher Ivo Kohler tested one subject who adapted to distorting prisms so well that he was eventually able to ride a bicycle through town. Halfway through the project, he experienced a curious phenomenon. "Inscriptions on buildings, or advertisements," Kohler wrote, "were still seen in mirror writing, but the objects containing them were seen in the correct locations. Vehicles [seen correctly] driving on the 'right'...carried license numbers in mirror writing."

How can the visual system, which seems so stable under ordinary conditions, possess the flexibility to adjust to such extraordinary distortions? The answer might lie in a sensory feedback mechanism linking the visual system with body movements. Exploring the reconstructed world through movement and touch seems to ease the adjustment. An experiment conducted by psychologist Richard Held of M.I.T. illustrates how the freedom to interact with the environment can help a human or animal adapt to visual distortion. Held raised two kittens in darkness for about two months, exposing them to light for

three hours each day. During this period, one of the kittens was permitted to walk around the inside of a cylinder with vertically striped walls. The other kitten sat in a gondola pulled by the first cat circling round the chamber. Although both saw the same stripes, only the active cat developed normal vision. The passive kitten remained visually impaired until it returned to a normal environment. Held attributes the difference to the first kitten's ability to correlate visual experiences with its own body movements — an opportunity the second kitten lacked.

If deliberate visual alterations reveal the flexibility of the visual system, distortions that arise unexpectedly show how vulnerable it is to internal influences. Hallucinations can distort vision alone or affect several senses at the same time. Most commonly caused by drugs, epilepsy or mental illness, hallucinations can also be provoked by sensory deprivation and migraine headaches. According to one theory, formulated by British neurologist Hughlings Jackson in 1931 and recently updated by Louis Jolyon West of UCLA, hallucinogens suppress the flow of information from the eyes and other senses, while the activity of brain cells remains high. As a result, says West, "images originating within the rooms of the brain may be perceived as though they came from outside the windows of the senses."

Under the influence of certain hallucinogens, floating impurities within the eye may become more visible. Sometimes objects may appear smaller or larger than normal or appear to change size as the viewer moves closer to them. This defiance of the law of size constancy perhaps occurs because drugs contract the ciliary zonule that controls accommodation. Dilated pupils, another side effect of drugs, might cause blurred vision, focusing problems and glare experienced by some drug users.

Three centuries ago, René Descartes postulated that pictures from the eye flashed onto a kind of screen inside the brain, where the soul sat and observed them. Modern scientists have unveiled a visual process that is far more complex and intriguing. The miracle of sight, once the brain's well-guarded secret, is rapidly yielding to man's understanding.

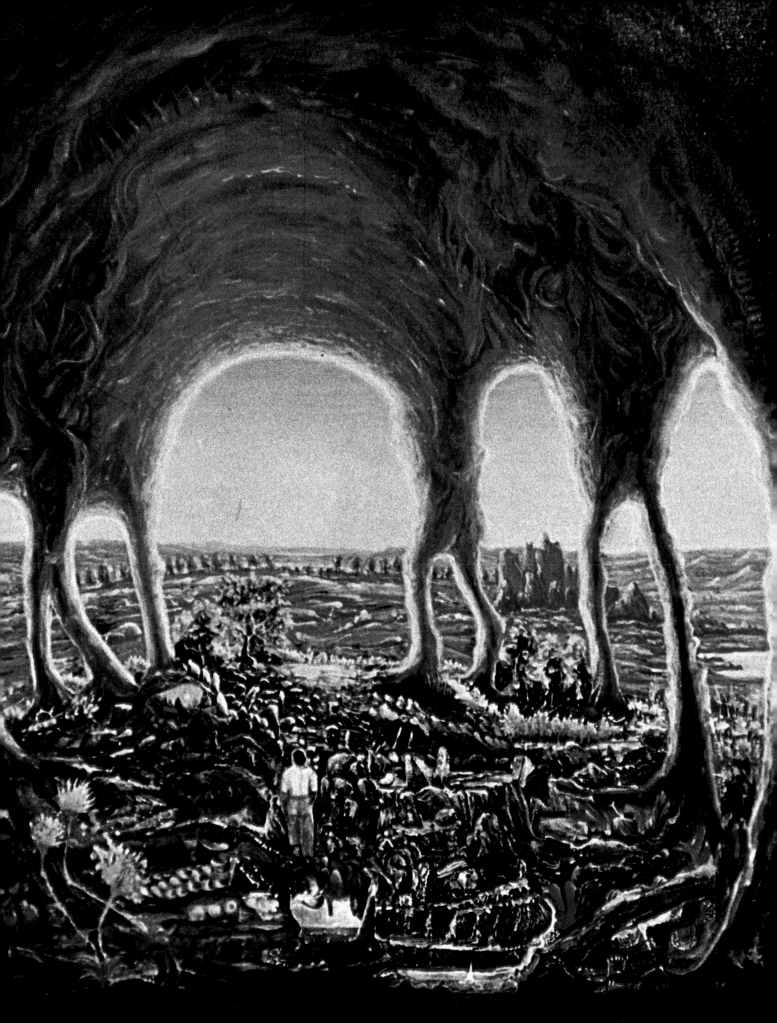

Chapter 4

Sensing Color

There is much that is certain in this multi-hued world. The sky is blue, the grass green and the sun a burning yellow. Leaves that bloom green in spring turn russet, brown and gold in fall. Winter brings snow that always falls white. By such color constants man knows and secures his place in the world.

Color, so seemingly tangible, is really anything but. Perhaps the purest abstraction, it is also the finest illusion. It does not belong to sky and sea and field. Color rises instead in the mind, a sensation triggered by the complex exchange between light and object, eye and brain.

Light is the sun's paintbrush, formed of radiant energy which the eye senses and converts to sight. The sun's radiation beams to the earth in an array of waves called the electromagnetic spectrum. The waves range in length from mere atmospheric ripples — gamma rays measuring no more than six-quadrillionths (.000000000000006) of an inch — to long, rolling radio waves that can stretch eighteen-and-a-half miles from one crest to the next.

The wavelengths that the human eye senses fall on such a narrow slice of the spectrum that their unit of measure is the nanometer — one-billionth of a meter. Only 300 nanometers separate light's longest wavelengths from its shortest. In a relative sense, man is all but blind.

About 400 nanometers long, the shortest wavelengths of light "fire" violet sensations. The perception changes to blue as waves lengthen, and from there to green approaching the spectrum's midpoint, 550 nanometers. The longer wavelengths, those ranging from 550 to 700 nanometers, kindle warmer sensations of yellow, orange and red.

In working the magic of color, light does not act alone. It needs objects and scenes to enliven. The physical world, able to absorb, reflect and refract light, thereby obliges, calling to mind the

Turning the spectrum inside out, Karl Gerstner pursues the spirit of color. Waves of yellow fade with subtle precision into deeper shades, spreading over a broad field of red. By the reflection of light off pigment, color opens into our eyes, giving beauty and meaning to all we see.

Light waves occupy a small segment
of the electromagnetic spectrum.
Measured by the nanometer — one-
billionth of a meter — each color
has its own frequency. Violet waves
are the shortest and red, the longest.

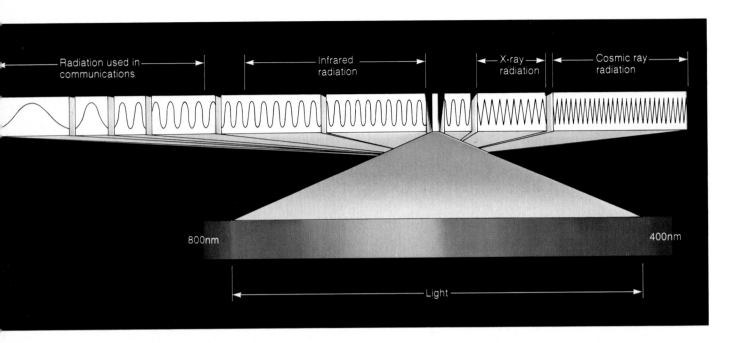

words of German poet Goethe: "Everything in life strives for color." Sensing the endless play of light on matter, the eye sends that experience, in code, to the brain, which breaks the code to reveal the wonder called color.

Almost every color sensation relies on the presence of pigment. Chemical, light-reactive substances that permeate the natural world, pigments vary in molecular structure. Each is thus equipped to absorb certain wavelengths of light and reflect all others. Pigment provides an organism with a means to convert energy to body heat while simultaneously regulating that intake to maintain proper body temperature. A by-product of this photochemical process is coloration, since the mix of wavelengths, reflected from pigment to eye, translates into color.

Man has extracted pigment from nature since the time he lived in caves and painted their walls to accent life. For ancient Egyptians, the making of pigments was a high art that predated — and rivaled in its ceremony — the practice of alchemy. They attempted to produce pigments synthetically but failed. It would not be until the nineteenth century, when chemical dyes were developed, that man could cease to rely on nature as his sole source of color.

From the earth the ancients drew somber browns, ochers and reds. They tapped veins of copper, lead and cobalt to yield dazzling mineral hues, while the vegetable kingdom offered up the brilliance of yellow, red and green. Madder root, dried and crushed to a powder, produced a deep red, while shellfish, drawn from the rivers and bays, distilled into airy violets and deeper purples. Sepia came from the dark fluid contained in the ink sac of the cuttlefish, cousin to the squid. Two species of snails found in the Mediterranean combined to make a royal purple. It is said that it took one-quarter of a million snails to produce one ounce of the prized dye.

For Greeks, Romans, Egyptians and ancient Chinese, the colors present in their art assumed largely symbolic value. Formal and bold, colors were often used to affirm the mystical beliefs of the culture. Watery blues or fiery reds expressed the elemental nature of existence. The artist subdued his inner vision and feelings to depict the culture's values.

Color suffered as well from technical constraint. Pigment was mixed with animal glues, waxes and egg yolk to make it bind to a surface. The product was a paint that did not work easily; applied flatly, it yielded images that lacked sub-

Sought for beauty and utility, pigments drawn from plants, minerals and animals have colored man's world for thousands of years. In some cultures, dyers have been accorded special status for their power to transmute natural objects into brilliant colors. Sumac berries, left, and copper, above, yield rich reddish brown dyes. The ten-armed cuttlefish, top, gives the name of its genus, *Sepia,* to the dusky hue distilled from the ink it secretes.

tlety and depth. Too, paintings did not age gracefully, for the medium, adhering poorly to the surface, would flake off with the passage of time.

Color as an element of art suddenly bloomed with a flourish in the fifteenth century. What brought it to life was not only the surge of the Renaissance spirit, but, more fundamentally, the improvement of technique. Oil-based paint, a new medium, freed both the hand and the mind of the artist. It could be applied to canvas in varying weight and worked quickly. It could be layered to produce a range of effect from transparency to opacity. Suddenly, with oil-based paint, art took on tone, clarity, shading. It could highlight not only the subject, but also the artistic impulse that brought the vision to life.

Oil as a "vehicle" for pigment emerged from tempera, or egg yolk, the best substance known for fixing pigment to canvas. Tempera was mixed with either oil or water. It was inevitable that artists would, in time, discover the virtue of oil. Who made this discovery is unknown. But if oil painting is said to have had "inventors," then perhaps they number two, the Flemish painters Jan and Hubert van Eyck. With this new medium, the van Eyck brothers produced works of unprecedented clarity.

But for such a technical milestone, details concerning the spread of oil painting are sketchy. Renaissance historian Giorgio Vasari credited a young Sicilian, Anotello de Messina, who traveled north to Flanders where he spent six years immersed in the study of the new medium. He returned to Venice and shared his newfound knowledge with fellow artists. So versatile did painters find oil that Vasari recorded: "Nothing else is needed save diligence and devotion, because the oil in itself softens and sweetens the colors and renders them more delicate and more easily blended." Anotello's journey proved to be a vital link between north and south, between the technical mastery of Flanders and the subsequent flowering of artistic genius in the Italian Renaissance. And perhaps there was no more formidable evidence of this fruition than the work of Leonardo da Vinci.

Leonardo studied the techniques brought back by Anotello — then set out to improve upon them. Before Leonardo put his mind to this task, the prevailing technique in oils was to mix pure colors with either black or white to yield shadow and highlight. But what the artist gained in form, he lost in tone and liveliness. Leonardo, in the development of his *chiaroscuro* style, kept his colors pure, achieving shadow and highlight through contrast. His premise was a simple one. Color in nature was pure, he decided, and that was how the eye saw it. So he painted what he saw. To soften and round the painting, he later applied glazing. The effect produced luminance in a scene, a brilliance that deepened the shadows and preserved the purity of his hues. Along with other giants of the Renaissance, Leonardo created and used color in ways that perhaps have not been rivaled since. Their genius, enhanced by the introduction of oil as a medium, ushered in the era of modern painting.

The "Proper" Color of Things

Color has intrigued and tested not only the artist, but the scientist as well. The essence of color, its nature and its purpose, has long been a subject of spirited scientific debate. Aristotle, generally considered the founder of modern science, sought to explain every natural occurrence and

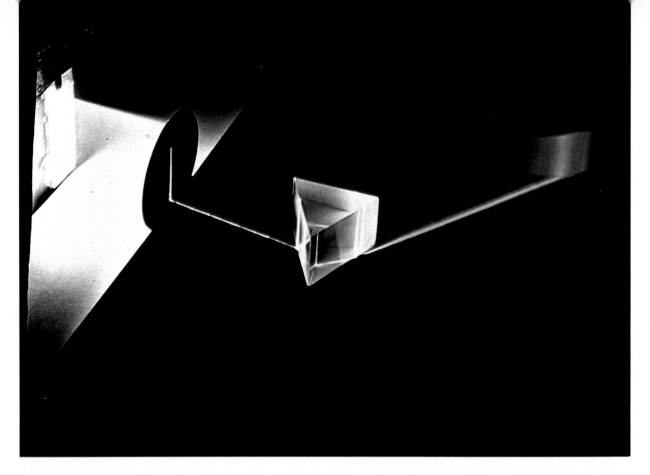

With a narrow opening in his shuttered window, Isaac Newton discovered how light separates into colors when passed through a glass prism. Summarizing his experiment, he declared, "it is manifest that the sun's light is a heterogeneous mixture of rays." In a modern reconstruction of his experiment, light directed into a prism bends according to the different lengths of waves within the spectrum, refracting in full color on white paper.

phenomenon his senses touched. Color did not escape his scrutiny. "Simple colors," he declared, "are the proper colors of the elements." All materials derived from the four elements — earth, air, fire and water. "Air and water when pure are by nature white," he reasoned, while "black is the proper color of elements in transmutation. The remaining colors, it may easily be seen, arise from blending by mixture of these primary colors." Thus, believed Aristotle, color must be a single quality bordered by the extremes of black and white. The gradual strengthening or lessening of one or the other could bring about the spectral hues. His was a theory that a simple experiment, perhaps never ventured by classical philosophers, could explode. Black mixed with white yields gray. Yet, Aristotle's view of color perception prevailed through the Middle Ages.

It was not until the seventeenth century, with the rise of the scientific method and the use of optics, that inquiry into color sharpened. Ultimately, the mind of one man, Isaac Newton, would hone it to a fine point.

Born a sick and only child in 1643, Newton was not expected to survive his first day of life, much less the eighty-four years he eventually lived. His father, a yeoman farmer, died three

months before Newton's birth. Remarrying two years later, his mother left the child with his grandmother for the next nine years. Neglected in childhood, Newton grew up a pensive loner who channeled his energies into pursuits of the mind — to the exclusion of those around him. Sent by his mother to tend the family's cattle, he would often settle under a tree, his head in a book, his thoughts far removed from the livestock's wanderings.

Fracturing Light

Newton arrived at Cambridge in 1661 when revolutionary scientific ideas were in the air but the doctrines of Aristotle were still in the curriculum. In 1543, Polish astronomer Copernicus had published *On the Revolutions of the Celestial Spheres*, a landmark work in which he postulated that man's existence was sun-centered. The Earth, said Copernicus, spun daily on its axis and revolved yearly around the sun. Sixty years later, German astronomer Johannes Kepler would describe the elliptical orbits of the planets and their spatial relationship around the sun. In France, meanwhile, philosopher René Descartes was busy overhauling man's conception of himself. Encouraged by a strong intellectual following, he saw not a system possessed with a life of its own but an impersonal machine operating by strict and impartial mechanical laws. The course that Newton charted through this sea of intellectual ferment is perhaps best revealed by a Latin phrase scrawled in a notebook reserved for traditional scholastic exercises. It translated: "Plato is my friend, Aristotle is my friend, but my best friend is truth."

Beginning in 1667, Newton lectured as a postgraduate fellow at Cambridge. He spoke on optics and the nature of light and color. Color, he said, was not an inherent, physical quality. It belonged not to the object, but to the mind. "For the rays to speak properly are not coloured," he declared, since "in them there is nothing else than a certain Power and Disposition to stir up a Sensation of this or that Colour." Furthermore, he continued, light was not simple and homogeneous, a single hue subject to modification, as Aristotle had thought. Rather, it was complex

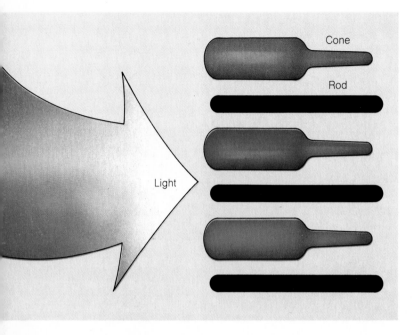

Cone

Rod

Light

and heterogeneous, "being everywhere mixed" yet made up of "pure," discernible colors.

Through experience, Newton offered proof for such a thesis. "In a very dark chamber" he opened "a round hole about one-third part of an inch broad" in the shutter of the window to send a beam of sunlight through a prism. The light fractured into the brilliant colors of the spectrum, then — beaming back through another prism — recombined into white light. This was an epic discovery born of a simple experiment. It would, however, remain largely unelaborated for 130 years before encountering the fertile mind of an English physician, Thomas Young.

Young took Newton's premise and — being a physician — focused his attention on the eye as an optical instrument. "It seems almost a truism to say that colour is a sensation," he reasoned. "The science of colour must therefore be regarded as essentially a mental science." But how could the eye, in relating color to the brain, possibly transmit every single shade of color that a scene could offer? Instead, Young thought it must have the power to simplify, to take basic data and channel it to the mind. Then the brain, more complex than the eye, could paint a visual masterpiece, faithful in every hue and detail.

Knowing through Newton's work that three primary colors, when mixed in varying number and strength, would yield almost any color, Young postulated that "the sensitive points of the retina" were "limited . . . to the three principal colours." These points — what modern scientists would later call photoreceptors — were tied to the brain, said Young. "Each sensitive filament of the nerve" thus consisted "of three portions, one for each principal colour." Each strand of the fiber could transmit a primary color, and the mind in turn could mix the sensations to yield color in all its possibilities.

Young's theory was at first rejected and later ignored. Scientific thought of the late eighteenth and early nineteenth centuries lay firmly in the grip of Newtonian principles. Young, a doctor, was probing color's physiological aspect, when the era — in homage to Newton — explained color by external, physical characteristics. It was not until half a century later that Hermann von Helmholtz, German physiologist and physicist, took up Young's theory and elaborated it.

Helmholtz at first rebutted the theory, believing that in some cases three primaries were insufficient to produce certain colors. After a decade of research, however, he changed his mind. His experiments finally suggested that color sensitivity in the eye was an overlapping phenomenon. Helmholtz believed that one wavelength of light could trigger more than one type of photoreceptor, producing two, if not three, primary color sensations of varying strength. These, when mixed in the mind, had the power to recreate any color the rainbow could offer.

Helmholtz revived Young's three-color idea in his monumental three-volume work, *Handbook of Physiological Optics*, which he wrote between 1856 and 1866. In establishing what was subsequently known as the Young-Helmholtz trichromatic theory, Helmholtz put forth an explanation of color vision that was both comprehensive and controversial. In staunch Teutonic opposition, Helmholtz's countryman and fellow physiologist, Ewald Hering, disagreed. He insisted the magic number was four, not three, and in 1870 he set before the scientific community his own ideas in the opponent-process theory.

Thomas Young

Challenging a Legend

When Thomas Young was fourteen years old, a woman visiting his tutor asked the boy, perhaps a little condescendingly, for a sample of his handwriting. Young complied by penning sentences in fourteen languages.

Even as a small child, Young knew how to make use of his remarkable talents. He could read by the age of two and by four had read the Bible twice. He became the first Western scholar to decipher Egyptian hieroglyphics. He studied and mastered medicine, chemistry and many other disciplines. Yet Young, who towered over his contemporaries, met fierce competition from the memory of Isaac Newton, a man who had died forty-six years before Young was born in 1773.

In the early nineteenth century, Sir Isaac remained England's greatest scientific treasure. But Young believed some of Newton's theories could be improved upon, a suggestion itself near heresy. In particular, he was convinced that Newton had been wrong to reject the wave, or undulatory, theory of light. His own studies of light persuaded him that light was a wavelike phenomenon.

Color vision was one of the mysteries of light that Young attempted to explain, a subject

virtually untouched by other scientists. Young measured the frequencies of wavelengths of colors along the spectrum. He found that red had the lowest frequency, violet the highest. Young, a physician, knew that the eye could not have a receptor for every color variation; indeed, the number of different color receptors must be quite small. After many experiments, he narrowed the number to three — red, green and violet. He theorized that man's perception of the total range of colors was built from varied stimulations of these receptors.

In 1801, when Young was twenty-eight years old, he was invited to deliver a series of lectures before the distinguished Royal Society of London. His subject was "The Theory of Light and Colours." Young knew his ideas would

not go uncriticized. He tried to soften the blow by using passages from Newton's works — albeit somewhat out of context — to support his wave theory of light. But Newton's scientific stranglehold on optics remained unbreakable. Shortly after Young's first lecture, an influential magazine, *The Edinburgh Review,* rose to Newton's defense.

"Let the Professor continue to amuse his audience with an endless variety of such harmless trifles," the magazine sputtered, "but in the name of science, let them not find admittance into that venerable repository which contains the works of Newton." Another of his lectures, said the magazine, contained still "more fancies, more blunders, . . . more gratuitous fictions, all upon the same field on which Newton trod, and all from the fertile, yet fruitless brain of the same eternal Dr. Young."

The attack worked. Young's beloved medical practice began to falter and he ceased his controversial theorizing. Yet he seemed to realize that he had already achieved enough. In 1802, he wrote, "The theory of light and colours, though it did not occupy a large portion of time, I conceive to be of more importance than all that I have done, or ever shall do besides."

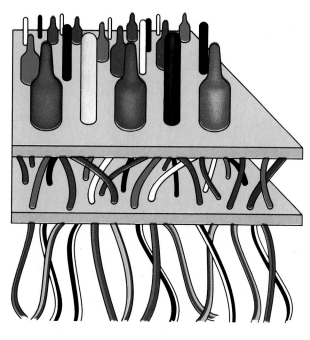

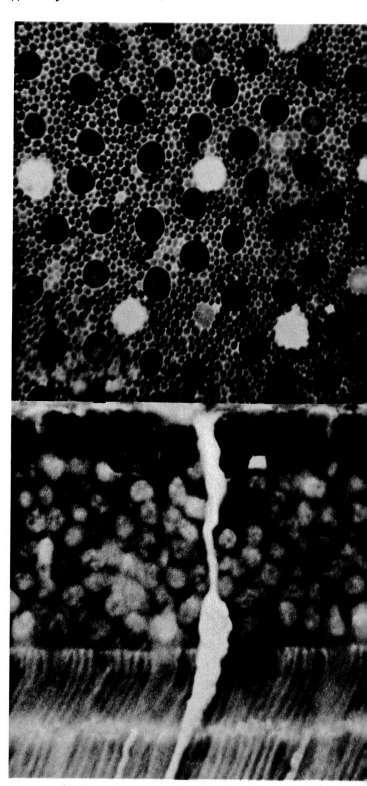

Hering discerned four primary colors — red, blue, green and yellow — as the psychological basis for all color sensation. These hues, noted Hering, seemed to arrange themselves in opposite pairs. Red and green light did not readily mix; there was no such color as reddish green. Instead, when blended, they fused into a distinctly separate color, yellow — the intervening primary hue. The same was true for blue and yellow. When mixed equally they yielded green.

Curiously, modern research seems to affirm that both the Young-Helmholtz three-color theory and Hering's four-color theory possess a measure of scientific truth, and actually complement each other. Taken together, they paint a reliable picture of what man currently knows about the mechanics of color vision. While the trichromatic theory serves to explain how the eye detects and deciphers color, Hering's ideas yield basic clues to the tangled mystery of how that same information is encoded and sent along the nerve pathways to the brain.

Research completed in the last two decades has confirmed that the eye indeed contains three types of color sensors, the photoreceptors lining the retina. Different photosensitive pigments enable each to absorb light primarily in the red, green or blue portion of the spectrum. These three types of photoreceptors are called red, green and blue cones.

Red cones respond to wavelengths of light measuring between 475 and 700 nanometers, with peak light absorption around 570. Green cones span the spectrum between 435 and 635 nanometers, peaking in light sensitivity around 535. Blue cones, which range between 400 and 550 nanometers, are most strongly stimulated by wavelengths that measure about 440 nanometers. Against this backdrop of spectral overlap, a wavelength measuring 510 nanometers would fire most strongly the green cones upon which it fell. Nearby red cones would be stimulated to a lesser extent, while blue cones would register the faintest spark.

Codes for Color Vision

The sensation of color begins when light strikes a portion of the retina, packed with the photoreceptors, rods and cones. Nerve cells sensitive to light, they in turn fire a flurry of impulses back through a maze of specialized intermediary cells, the bipolar, horizontal and amacrine cells. These signals cross yet another gap to reach ganglion cells, the final retinal link to the optic nerve, eye's pathway to brain.

While rods translate the power of light into the monochromatic values of black and white, cones absorb light and encode it for color. In keeping with Hering's opponent-process theory, they give each stimulus strength relative to the opponent color sensation, thereby creating a specific ratio of "on-off" signals. Impulses transmitted by red cones tend to turn on the sensation of red through the fibers that tie them to the optic nerve. Those same impulses simultaneously tend to inhibit, or turn off, the brain's ability to see green. The exact reverse is true for impulses that excite green cones. Similarly, impulses sent through blue cones oppose the combination of signals sent by red and green cones — which together produce yellow, blue's opponent color. By this theory, a tree full of limes strongly excites green cones and their connected ganglion cells. Such a vision has an equally inhibitory effect on the brain's ability to see the opponent color, red.

Ultimately, the brain absorbs this ever-shifting flood of impulses, weighs it, synthesizes the discrepancies and thereby builds a precisely painted picture of what the eye has sensed. The opponent process enables the mind to take a wealth of generalized color information, funnel it through the retina's five million cones, and hone it down to a pinpoint of color. The result is a mosaic crafted by a process in which each color sharpens and thus defines its opponent.

The stream of images from world to mind weaves seamless patterns of color at the speed of light. That the visual system works so adroitly, so reliably, is a source of awe. Only in about 5 percent of the American population does color vision go awry, leading to colorblindness.

Colorblindness is a misnomer, since in nearly every case there is a deficiency, not a total lack, in the eye's power to sense color. This deficiency stems from two principal causes. Either certain types of cones — red, green or blue — are absent from the retina, or the information they gather is improperly or incompletely conveyed.

True colorblindness, monochromatic vision, is a statistical rarity. It occurs in about one in every 40,000 people. People with this condition are called rod monochromats, for only their rods function. They see in shades of black and white, a world of endless dusk. Even rarer than rod monochromacy is a condition in which only one type of cone is present, tinting a person's total visual experience with either red, blue or green.

Far more common is dichromacy, a condition in which the retina lacks one type of cone. Absence of blue cones is very rare. But the lack of either red (protanopia) or green (deuteranopia) cones accounts for the major color defect — red-green colorblindness. Together, they afflict about 4 percent of the American population.

If the eye lacks red cones, light falling in the red portion of the spectrum stimulates only green cones. This skews the normal ratio of excitatory and inhibitory impulses between red and green cones, creating an abnormally strong green sensation. The reverse essentially occurs when the green cones are missing, with wavelengths striking the green portion of the spectrum deflected to either the red or blue segments of the spectrum.

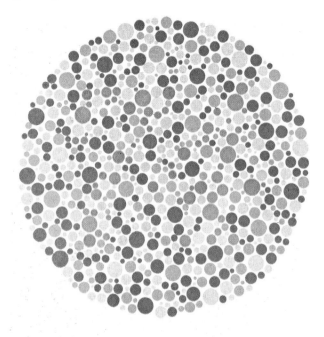

Dichromacy can also occur when the red and green nerve channels fuse and so produce the sensation of yellow, with the subsequent loss of red-green opponency. Faulty neurological connections between eye and brain or the mixing of red and green pigments in the same cones can bring on this condition.

In a third form of colorblindness, anomalous trichromacy, all three types of cones are present but working in disproportion. The information they yield is "shifted" on the spectrum away from peak norms. In anomalous trichromats, the perception of "pure" reds, greens or blues is literally a shade off normal response levels, leading to diminished sensitivity to part of the spectrum. An anomalous trichromat might see bright red as orange, or blue as violet.

Colorblindness is hereditary, a trait bound up in the twenty-three pairs of chromosomes that pack the body's every cell. Chromosomes carry the chemical clues that give shape, bearing and character to the human form. Between the male and the female, only one of these twenty-three pairs, the sex chromosomes, differs — and herein resides the secret of colorblindness.

In the female the two sex chromosomes are essentially identical. Geneticists, identifying them by their approximate shape, call them X chromosomes. In the male, however, the two do not match. One resembles the female's X chromosome, but the other is smaller and shaped differently. It is called the Y chromosome. The paired female sex chromosomes take the designation XX, and the male's, XY.

When the gametes, primitive sex cells, divide in the process called meiosis, each new cell takes half the pair of chromosomes, including one of the two sex chromosomes. Meiosis creates two types of sperm cells, one carrying an X chromosome and one, a Y. In females, it produces only one type of ovum bearing an X chromosome. When sperm and ovum meet, they merge two X chromosomes to produce a female offspring, or an X and a Y, which yield a male.

The genetic message for faultless color vision rides on the X chromosome. It is also a dominant trait, which means it need only appear on one X chromosome. Females, with two X chromosomes, are statistically far less liable to inherit colorblindness than males. The affliction in females, requiring the match of two color-defective X chromosomes, can only occur when both parents have at least one color-defective X chromosome. For males, though, only the mother need have one color-defective X chromosome to pass the deficiency along. While colorblindness occurs in about 8 percent of the American male population, its appearance in females is roughly .64 percent, or sixty-four out of ten thousand women.

Although colorblindness might sap the eye of some of its perceptual strength, it cannot alter the essential nature of vision. The eye is a tiny but abundant organ that forges man's major link with the world. Through it he gathers a flood of fact, mirage and nuance — visual riches that bear the wealth of the rainbow.

Color, so strong and precise, is also a master of illusion. The sky is not blue. What we see is a result of the atmosphere's scattering, or deflecting, short "blue" wavelengths of light eleven times more effectively than those that fall further along the spectrum. While the artist mixes the primary hues of blue, red and yellow to make all others, the physicist chooses blue, red and green as his primary colors. The key to this apparent

The artist can never perfectly duplicate the effects of light because pigments and light do not combine in the same way to form colors. Paints absorb light, reflecting fewer waves back to the eye. When the artist mixes his primary hues, a murky black results. Light waves, though, add to one another when mixed, making more of the spectrum visible. The primary colors of light combine to form white.

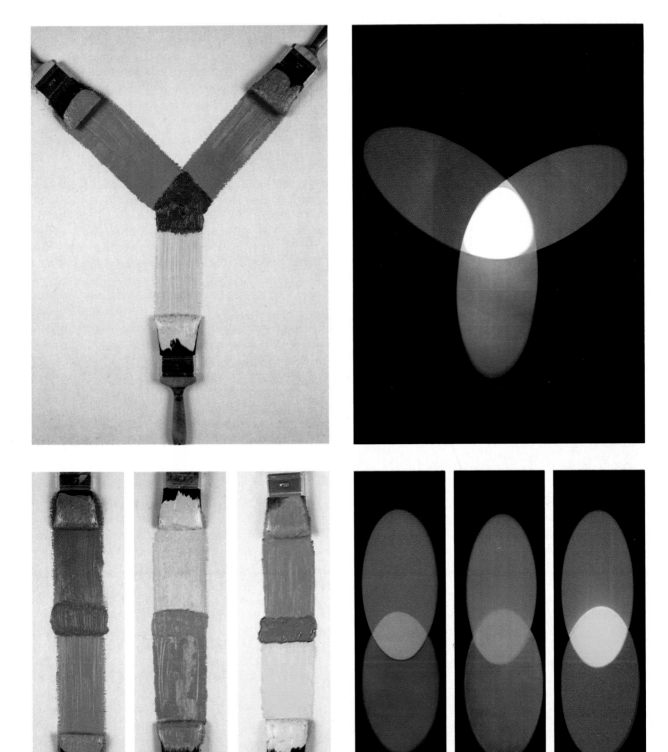

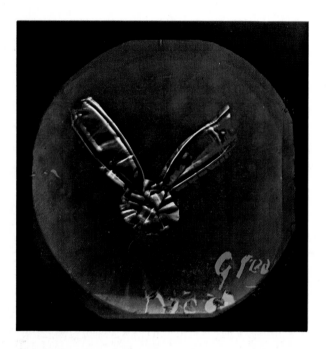

The first color photograph ever made, James Maxwell's 1861 image of a ribbon substantiated his belief that all colors were combinations of red, blue and green. A Scottish physicist, Maxwell took a black-and-white photograph of the ribbon and then projected a slide made from the negative through tinted filters onto a white surface to give the effect of natural color. His technique, the basis of color photography for 75 years, relied on the additive principle of mixing colors.

contradiction lies in the differing natures of pigment and light. When mixed, pigments subtract wavelengths. They absorb more wavelengths, leaving less of the spectrum to be reflected back to the eye. Beams of light, when mixed, have the opposite effect. They add wavelengths together, combining more of the spectrum for the eye to sense. In this way, two pigments on a palette and two beams of light, though identical, will not always yield the same hue when combined, for the mixtures of pigment and light offer the eye different wavelength combinations.

In the mixture of sun and rain, color achieves perhaps its most illusory effect. Arched on the stormy horizon, a rainbow appears when the sun drifts from behind a cloud to use a curtain of afternoon rain as both mirror and prism. Sunlight, shot into the shower and refracted, or bent, by the denser watery medium, is then reflected back out and scattered into spectral splendor.

The rainbow has perennially intrigued both artist and poet, but it was French philosopher René Descartes who first cracked its mathematical essence. Isolating a drop of water, he traced a number of the sun's rays through it and found a concentration of bent and reflected rays emerging at one angle. That concentration of exiting rays intersected the entering rays — or the sun's path — at an angle of 138 degrees. Because of the drop's tendency to concentrate light within the small interval of about one degree, the emergent light at this point does not scatter. It bends cleanly, fragmenting into different wavelengths. This is the rainbow.

The rainbow's shower of light always describes a 138-degree angle with the course of the sun's rays. For this reason, the rainbow appears either early or late in the day. When the sun rises higher than 42 degrees off the horizon, the 180-degree angle of the earth's horizon cannot accommodate the vital 138-degree angle that gives the rain its coat of many colors. Similarly, the lower the sun sinks the higher the rainbow rises.

Color often tricks the eye, and so the mind. Man's desire is to know it, to define its essence. Science sees color as having a three-dimensional "shape," a form built around three factors called hue, brightness and saturation.

Whether seen directly or through
the diffraction of lenses, light clothes
the world in a stunning wardrobe of
color. Raindrops separate rays of
light into the different wavelengths
that make up the colors of the
rainbow, perhaps the most wondrous
effect of light in nature, previous
page. As the angle of the sun falls
in the evening, light strikes tiny
particles in the air, tinting the
western horizon with the roseate
glow of sunset, above. Seen under a
microscope, the grooved cones of
vitamin C assume spectacular hues
when exposed to polarized light,
right. Light reflects at different
angles from oil and from the layer of
water beneath it to form a glistening, abstract spectrum, opposite.

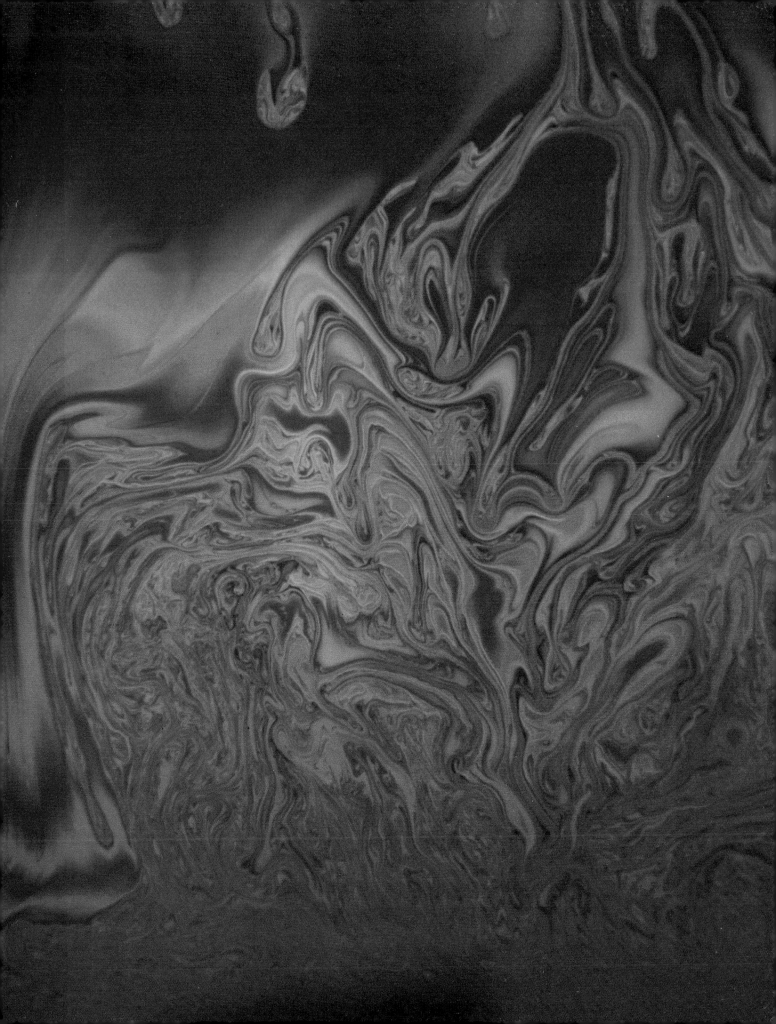

Hue is a color's name. Like cardinal points to the compass, four primary hues — those identified by Hering — anchor and give shape to the world of color. Arrayed in a circle, or color wheel, red, green, blue and yellow are seen as the simplest colors; they appear as hues that have not been mixed from other colors. Like north and south, east and west, red and green and yellow and blue forever oppose each other; they never merge. At the point on the color wheel where two primaries meet in equal strength, the result is not a blend but a fusion — a distinctly different hue in the form of another primary. Blue mixed with yellow becomes not bluish yellow, but green. Red and green fuse into yellow.

Brightness, or luminance, rises from the amount of light energy striking the retina. Though the greater the amplitude, or height of the wave, the brighter the sensation, luminance is tempered by wavelength. Because the eye is more sensitive to light in the spectrum's center, medium-length wave sensations — oranges, yellows, pale greens — seem brighter than reds and blues of equal energy, which appear, respectively, on the long and short ends of the spectrum.

Saturation describes a color's purity, the degree it is free of whiteness or blackness. Saturation is the product of a light wave's simplicity. A light wave composed of only one wavelength paints the richest impression of saturation. As that light wave grows more complex with differing wavelengths, it dilutes and drains purity from the hue. Curiously, allowing for uniform luminance, the brighest hue, yellow, also appears the least saturated; while the relatively less luminous reds and blus on the spectrum's edges appear richer.

The shape of color, when expressed as a mathematical model, is a pair of cones joined base to base and aligned on a vertical axis. The circumference of this geometric solid at its widest point describes the color wheel, the array of hues in their purest state. Saturation covers the circular plane which joins the two cones. It radiates from a gray core to a pure periphery, and there it meets the color wheel. Brightness travels the model's vertical axis, rising to a peak of ultimate brightness, whte, and dropping in the other direction toward black. Within this model, the average eye can sense seven million colors.

But the eye alone does not perceive color. The source of color lies in the intricate meeting of worl and mind, the interplay of context, contrast, memory and mood. Hering believed the perception of color was based on no less than "the optical experiences of a whole lifetime."

Space, Time and Color

In the sunless chill of a gray November day the land unfolds through the fields of winter to meet the horizon. In the middle distance stand trees, dark and skeletal against a sky the color of steel. It is a scene asserting the season's inevitability. And yet that season is not forever. The clouds will part to reveal a sky of cobalt blue, the frozen earth will warm, and the trees will bloom.

Such sharp changes in nature are elaborate variations of the phenomenon of contrast. In spatial contrast, the relative position, size and brightness of objects in a scene create subtle shifts in the perception of hue. A gray square on a red background appears slightly greenish, while on a green background, it tends subtly toward red. A dark object in the distance appears bluish, a brighter object, orange or reddish. Two objects, one bright and the other dark, usually seem even

94

Brightness

Saturation

Hue

The varying shades of color can be charted on a three-dimensional model. Hue, brightness and saturation — color's three qualities — multiply in orderly infinity when they inevitably meet. Hue is the name of a color, and brightness is the amount of light reaching the retina. The purity of a color's tone is known as saturation. Hues are purest at the outer edge of the central horizontal plane, far from the diluting effects of gray, which appears at the model's core.

Proclaiming the autonomy of color, Josef Albers used a flat, ever-constant pictorial format in his Homage to the Square series. Albers demonstrated the complexity of perception by varying only the colors in a pattern of overlapping squares. One square deceptively looms larger as the interaction of colors pushes it outward. By the manipulation of the artist, colors move forward or fade into the background, and appear thin or dense.

more so when juxtaposed. If separated, they grow more alike in tone. Wide and endless variations in spatial contrast can be produced, scientists think, because of the fine mesh of neurons that unite all parts of the photosensitive retina. The eye, sensing different colors all through a scene, trades and compares that information through the opponent process. Thus, a color, seen on the periphery of a scene, can serve to soften or highlight a second color apearing at the center.

It was Helmholtz who was the first to note that some colors can appear brighter than others at higher saturations, even though they are equally luminant. Friedrich Kohlrausch, fellow German physicist and student of light, elaborated on this effect. He placed two colors of equal luminance but different saturation side by side. Then he flickered the lighting so that the patches appeared alternately. The perception Helmholtz had noted disappeared and the two patches seemed of equal brightness.

This phenomenon, the Helmholtz-Kohlrausch effect, implied that contrast is not only spatial but occurs through time as well. In this case, time intervenes to erase the perceptual imbalance created by contrast, so the brain perceives each color separately. A simple example of temporal contrast is the illusion called a negative afterimage, which arises when the eyes stare for a period of time at a patch of color. When next presented with a white background, the eyes will see the opponent color. A red patch stimulates a green afterimage, and vice versa. The longer the view of the first color, the more lasting the afterimage.

Though spatial and temporal contrast lack hard scientific explanation, they appear to support Hering's opponent-process theory, which implies that color vision, like other natural systems, strives for balance. In the case of color vision, the search for balance takes place in the mind.

The more opposite two colors the more they fire the extremes of inhibition and excitation in the retina's response. Likewise, the concept of balance in color vision teeters through time. The longer the eye stares at a red background the more the perceptual balance tilts away from green. When freed from seeing red, the eye, its red cones overstimulated, will yearn to see a greenish hue of comparable strength and duration. In this way it rights the balance.

Color ultimately takes its form from more than a series of passing perceptions. Its roots reach back in time to when man clung to richly colored myths and symbols to explain and order existence. Red is the color of passion and life. It has been so ever since man understood the vitality of fire and blood. Green represents fertility, a connection as eternal as the annual coming of spring. Black is the darkness of death, and white, the purity of noble spirit. The task of interpreting color falls to the painter. His intent surpasses the faithful transfer of image and scene. He seeks, as well, to merge moment and myth.

Square of Blue, Streak of Yellow

"When you go out to paint," advised Claude Monet, master of French impressionism, "try to forget the objects you have before you, a tree, a house, a field.... Merely think, here is a little square of blue, here an oblong of pink, here a streak of yellow." Light and color, Monet believed, were a painting's essence, the very soul of art. The critics of the time thought otherwise.

That color took precedence over a painting's subject proved a revolutionary idea in the world of late nineteenth-century European art. The idea grew into a movement, impressionism, taking its name from an early Monet work, *Impression at Sunrise*. In 1874, the painting hung in a Paris show organized by Monet and a handful of other painters who shared his vision. The critics came and saw works they called "incomplete." What they were looking at was art that celebrated the medium — color — and therefore deemphasized the long sacred elements of object and detail.

Monet was engaged in a pursuit as consuming as Newton's, two centuries earlier. He sought, with paint, to break light down and to crack it open. Scientists of his era were beginning to look at light not only as form and medium, but as energy. The development of the camera also gave man the technical means to study the mood of light from moment to moment. Art, borrowing from science, reaped the aesthetic rewards. Man's newfound ability to chemically synthesize and mass-produce pigment made painting a portable

Fascinated by light, Claude Monet painted it as a pattern of tiny nuggets of color, changing with the time of day. His haystack paintings of 1891 show a subdued sunset and the glint of light on snow.

pursuit. It freed the artist to move out of the studio's muted light and into nature's brilliance.

The impressionist palette similarly brightened. Earth tones — umbers, siennas, ochers — gave way to the brighter primary hues. The intensity of Monet's palette mirrored his pursuit. French writer Guy de Maupassant, on observing Monet at work in 1885, recalled:

> He was no longer a painter, in truth, but a hunter. He proceeded, followed by children who carried his canvases, five or six canvases representing the same subject at different times of day.... Before his subject, the painter lay in wait for the sun and shadows, capturing in a few brush strokes the ray that fell or the cloud that passed.

For Monet, the hunt narrowed with age. From the mid-1880s until his death in 1926, he painted scores of the same scene, day in and day out. Through these "series" he sought the "transposition of nature, at once forceful and sensitive."

But the task often proved greater than he thought. He set out to paint one landscape with two canvases, one for the morning, one for the afternoon. Soon he had filled forty. Sometimes he would spend only fifteen minutes on a canvas before putting it aside because the light had shifted on him. In 1892, he rented an apartment on the city square in Rouen, across from the cathedral. Over the next two-and-a-half years he would paint its stone facade, ever-changing with the passage of light and time.

Toward the end of his life, Monet began losing his sight to cataracts. He despaired, but kept on working, going out to paint at dawn and dusk when the softer light would not hurt his eyes so much. His search was impossible, and he knew it. Monet was obsessed with painting the moment, rendering a faithful portrait of nature, a subject that would never sit still. He seemed mindful of the paradox when late in life he wrote: "I have only the merit of having painted directly from nature, trying to convey my impressions in the presence of the most fugitive effects." It was, perhaps, an appropriate epitaph for the painter as well as an apt intimation of color's beguiling nature. Color, such a pillar of perception, also gives life its fancy. It is nature's most fugitive effect.

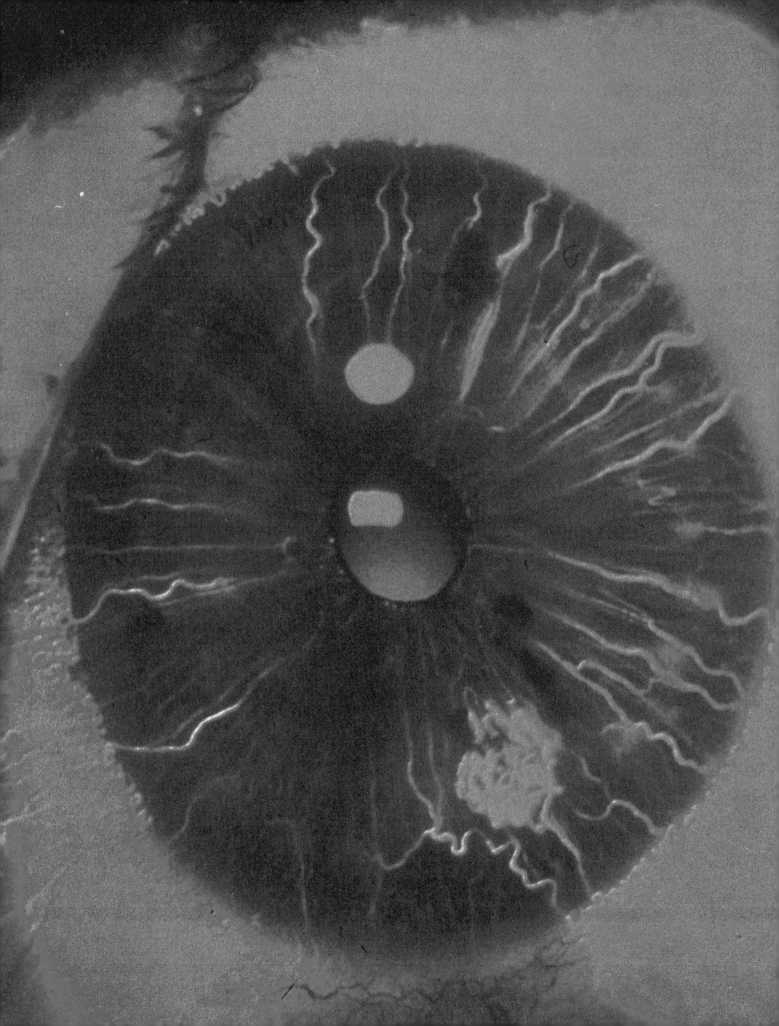

Chapter 5

The Imperfect Eye

Like any camera, the eye aims, focuses and adjusts for changes in lighting. Like any organ, it grows, labors, tires and ages. Ophthalmology demands a curious shuttling between studying the eye as a focuser of light and as an organ of the human body. How well the eye sees helps to reveal how well it is. As the eye yields clues about its own health, it also hints at the well-being of the whole body. The eye is the body's only transparent organ, a window out and a window in. Brain tumors, high blood pressure, bacterial infections, multiple sclerosis, even drug abuse can all leave a trace in the eye. More than a century ago, English physician Peter Mayer Latham advised his students, "In the eye you will see all diseases in miniature, and you will see them as through a glass."

The most familiar eye examination tests visual acuity — how well the eye detects patterns at various distances. When a patient reads the rows of letters on a wall chart dominated by a huge, capital "E," he is demonstrating his visual acuity. The chart hanging on the walls of most ophthalmologists' offices today was created by Hermann Snellen, a nineteenth-century Dutch ophthalmologist. Snellen used the acute eyesight of his assistant to develop a standard system for testing all eyes. By measuring the limits of his assistant's vision, Snellen devised a chart with letters of different sizes to represent what the normal human eye should be able to see at 20 feet, 30 feet and up to 200 feet. A patient who can read the row of letters that Snellen's assistant could read at 20 feet has 20/20 vision. But if the first row he can read represents what the normal eye should see at 100 feet, the patient's visual acuity is 20/100. Although other methods of measuring visual acuity have appeared since 1864, the year he introduced his test, the Snellen chart remains the most widely used. Along with the chart, many ophthalmologists use a near-vision test to exam-

Green rivers of fluorescein dye fill the tiny blood vessels of the iris. The twisted mass of blood vessels on the right marks a tumor. Fluorescein glows a bright yellow green under blue light. By injecting the dye into the blood stream and taking a rapid series of photographs as the dye courses through the vessels of the eye, scientists can create a fluorescein angiogram, a permanent record of the eye's health or disease.

101

An applanation tonometer pressed gently on the cornea measures the intraocular pressure — the pressure of fluids inside the eyeball. High intraocular pressure is one of the symptoms of glaucoma.

The beam from a slit lamp illuminates the landscape of a human eye, the ridges and gullies of the iris radiating out from the central black crater of the pupil. A narrow slit and a condensing lens shape the beam of a slit lamp into a brilliant rectangle of light. By directing the beam at different parts of the eye and looking through a biomicroscope, ophthalmologists can examine the cornea, iris, lens, vitreous humor and other structures.

ine a patient's eyesight at shorter distances. Like a Snellen chart in miniature, the near-vision test measures visual acuity from fourteen inches to seven or eight feet.

By using a pinhole test, an ophthalmologist can partially determine whether poor vision is the result of an error of refraction, such as near-sightedness, or a disease. The patient looks at a chart or object through a pinhole in an opaque card. The pinhole only lets in the rays of light streaming directly from the object to the center of the patient's cornea. Few of these light rays are refracted, or bent. They travel straight down the visual axis of the eye to the retina and thus eliminate errors of refraction that might appear if more light entered the eye. The vision of a healthy eye should be limited but clear. If the pinhole test does not improve the patient's vision — or makes it worse — the ophthalmologist suspects disease.

Inside the Eye

The ophthalmologist uses other simple tests to check color vision, eye movements, peripheral vision and depth perception. But for the most important part of an eye examination, the physician must look inside the eye. Using a slit lamp and a biomicroscope, the doctor can examine the structures of the eye. Chin and forehead rests hold the patient's head in place while the ophthalmologist shines a narrow beam, like a slice of light, into the eye. Peering through the biomicroscope, the doctor sees reflections of the eye's inner world. A slit-lamp examination can reveal tiny scars and ulcers on the cornea, damaged cells in the fluid of the anterior chamber, cloudy spots in the lens and floating particles or blood in the vitreous chamber. Special lenses permit the physician to see through the vitreous and observe the retina. After attaching a device called an applanation tonometer to the slit lamp and anesthetizing the eye, the physician pushes on the surface of the cornea to determine the pressure of fluids inside the eye, a crucial test for diagnosing glaucoma.

The most revealing path into the eye is through the ophthalmoscope. A simple device, the ophthalmoscope consists of a small lamp that beams a narrow ray of light into the eye, a view-

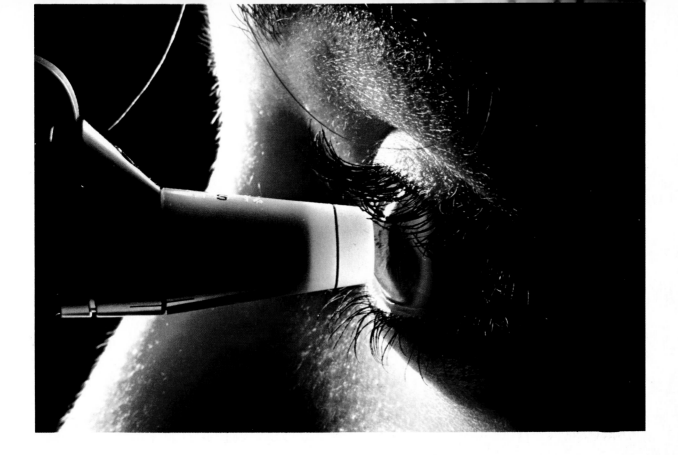

ing aperture and a wheel of lenses that rotates in front of the aperture. By aiming the illuminated ophthalmoscope at the patient's eye and peering through the opening in its center, the doctor puts his eye directly in the line of light reflected from the interior of the patient's eye. Rotating the lenses on the wheel enables a doctor to focus on various structures, including the cornea, lens and retina. Without using eye drops to draw back the patient's iris to dilate the pupil, the doctor can see about 15 percent of the retina. A dilated pupil reveals more than half of the eye's interior. The ophthalmoscope magnifies the image of the eye's inner structures about fifteen times. An indirect ophthalmoscope, consisting of a brighter lamp, a binocular headpiece and a series of hand-held lenses, reveals the entire retina in three dimensions, magnified two to five times. This three-dimensional view is invaluable in diagnosing glaucoma, torn retinas, swelling of the optic nerve or small tumors in the eye. The slit-lamp examination and the ophthalmoscopes reveal a strange, tiny world, the pale background of the retina covered with a tracery of blood vessels, like red rivers on an orange earth. Every spot, patch, bulge, shadow and unusual coloration of the retina could signal eye disease.

Using a green dye called fluorescein, a physician can check for nicks and ulcers of the cornea. Under blue light, the dye glows a bright yellow green. Washed with the dye and illuminated, a scratched cornea shows a distinct fluorescent stain. Ophthalmologists use a fluorescein wash to check contact lens wearers for scratches of the cornea. Left in too long, contact lenses can damage the cornea's delicate surface.

Optical Flaws

The problems doctors discover most frequently are not diseases but refractive errors. The four common refractive errors — nearsightedness, farsightedness, astigmatism and presbyopia — are seldom signs of deteriorating vision or disease. They are usually minor flaws in the construction of the eye. The human eye attains its adult size and shape when a person reaches the twenties. At that point, nearsightedness, astigmatism and farsightedness usually stabilize.

Nearsightedness, known scientifically as myopia, means that a person sees nearby objects more clearly than distant objects. It does not mean that a myopic person has excellent near vision and poor distance vision. Extremely nearsighted people do not see anything clearly. In the normal

Nearsightedness

The nearsighted eye focuses rays of light before they reach the retina; from the focal point, the rays begin to diverge, resulting in a blurred image.

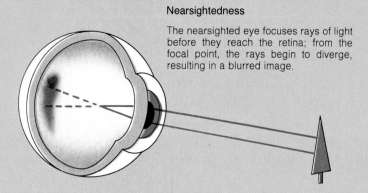

Concave lens

A concave lens spreads rays of light slightly before they reach the cornea, correcting nearsightedness.

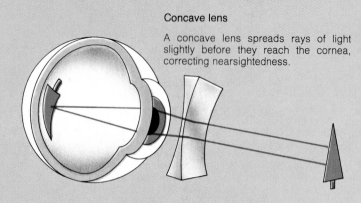

Farsightedness

The farsighted eye does not refract rays of light sharply enough to bring them into focus on the retina.

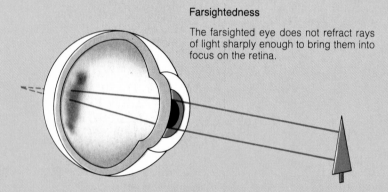

Convex lens

A convex lens bends rays of light inward before they reach the eye, correcting farsightedness.

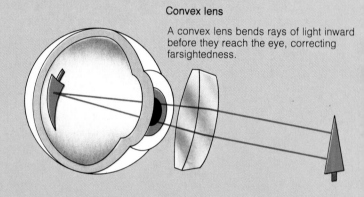

Astigmatism

The irregular shape of the astigmatic cornea bends rays of light unevenly, leaving a blurred image on the retina.

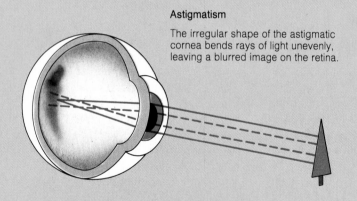

Cylindrical lens

A cylindrical lens bends only certain light rays, compensating for the oval shape of the astigmatic cornea.

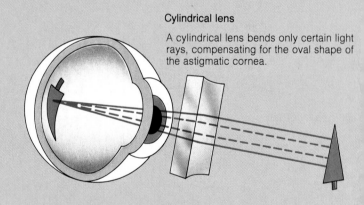

Presbyopia

As the eye ages, it loses much of its power to bring near objects into focus by changing the shape of its lens, a process called accommodation.

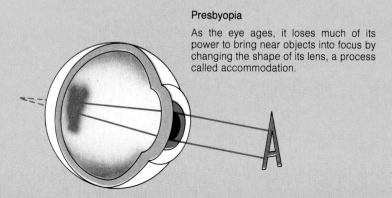

Bifocals

A convex lens, often in the bottom of bifocals, bends light inward and helps the presbyopic eye focus on near objects, like the print on a page.

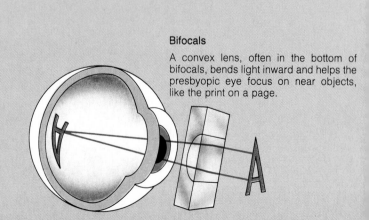

*Nearsightedness, farsightedness,
astigmatism and presbyopia are the
four common refractive errors.
About half the people in the U.S.
wear glasses or contact lenses to
correct these optical flaws.*

eye, parallel rays of light passing through the cornea bend slightly and then bend again as they penetrate the lens to focus on a small, single spot on the retina. In the nearsighted eye, a steeply curved cornea or an elongated eyeball, or both, focus rays of light before they reach the retina. From this focal point the light rays begin to diverge. By the time light reaches the retina, the image is blurred.

Farsightedness, known as hyperopia, is the opposite of myopia. It is not necessarily the ability to see objects clearly at a distance and poorly close up, although this can be the case. In the farsighted eye, rays of light streaming through the cornea and lens reach the retina before they come into focus, producing a blurred image. In a sense, the eyeball is too short for the refractive power of its own optical system. Farsighted children can bring images at almost any distance into focus by using the powerful ability of their eyes to accommodate. Accommodation is the process of using muscles of the eye to change the shape of the lens and increase its refractive power. But the primary role of accommodation is to help bring near objects into focus. By accommodating too often, farsighted children may strain the machinery of their eyes. Instead of suffering blurred vision, a few get headaches and eyestrain from overused muscles. Sometimes, mildly farsighted people do not discover the refractive error until middle age, when their powers of accommodation decline and they begin to have trouble seeing clearly at close distances.

An astigmatic eye results from an imperfectly shaped cornea. The spherical surface of a normal cornea bends light equidistantly. Rays of light entering the normal eye vertically and others entering horizontally refract equally and merge to a focal point on the retina. But the cornea of the astigmatic eye might be poorly curved from top to bottom or from side to side. This error of design gives the cornea different refractory powers across its length and width and distorts sight. Some light rays focus in front of the retina and others, behind it. As a result, the retina never receives a single, sharp image.

Presbyopia, meaning "old man's eyes," is an optical flaw that eventually affects almost every-

one. As the eyes age, their powers of accommodation decline. It becomes increasingly difficult for the eye to bring nearby objects into focus. Because they cannot see nearby objects clearly, some people confuse presbyopia with farsightedness. But presbyopia and farsightedness are separate conditions. Indeed, an individual with presbyopia can also be farsighted, astigmatic or nearsighted at the same time. Presbyopia, in a sense, gave bifocals to the world. The lower lenses of bifocals are designed to give presbyopic eyes extra help with nearby objects, no matter what other refractive errors the eyes may have.

Aids to Vision

Lenses of different shapes — convex, concave, cylindrical or some combination — can correct all four major refractive errors. Neither glasses nor contact lenses, however, can change a flaw in the eye; they only correct vision.

To prescribe a patient's first pair of glasses, the physician uses a retinoscope to measure the patient's refractive errors. The retinoscope beams a narrow stream of light onto the cornea. By placing lenses between the patient's eyes and his own, the doctor can use the shape and movements of the reflections to determine how well any combination of lenses corrects the patient's vision. Looking through various lenses, the patient reads from an eye chart to confirm the doctor's diagnosis. To change a prescription, the ophthalmologist checks the old pair of glasses and then asks the patient to look through various lenses to determine which provide the clearest vision. Ophthalmologists use either a trial frame in which lenses are inserted and switched or a phoropter, a wheel of lenses that rotates in front of the patient's eyes.

The power of lenses is measured in diopters. A lens's focal length is determined by measuring the distance between its center and the point at which it focuses parallel rays of light. To obtain the strength of the lens in diopters, the focal length is divided into 100. If the focal length of a lens is 50 centimeters, its power in diopters is 2 — 100 divided by 50 is 2. A lens with a focal length of only 2 centimeters is a 50 diopter lens. The shorter the focal length, the greater the strength

VIRGILIVS.

Vtuid Cuma... oenr iam... gon nis
ætas Maguus ab integro feclorum uafeitur ordo.

Soon after the invention of spectacles in the late thirteenth century, some painters put them in portraits of classical and Biblical figures. In this Renaissance portrait, an unknown artist has dressed Roman poet Virgil in the robes of a scholar, given him leatherbound books to read and a pair of spectacles to aid his aged eyes. Virgil, author of the Aeneid, *lived in the first century* B.C. *The robes, books and spectacles tell more about the painter's era than Virgil's.*

measured in diopters, and the more powerful the lens. Convex lenses bring parallel rays of light together and are called "plus" lenses. Concave lenses, because they spread parallel rays of light, are "minus" lenses. Farsighted people need convex lenses to draw rays of light together on the retina. Concave lenses spread light rays to compensate for the overly strong optical system of the nearsighted eye. A cylindrical lens refracts rays of light in one plane and can be ground so that it bends only those rays that lie along the weakest axis of the astigmatic eye. No matter how near, far or astigmatically sighted a person may be, a convex lens is necessary to aid presbyopic eyes, if only for reading. In many cases, a single pair of glasses or contact lenses combines the properties of different types of lenses. An astigmatic, nearsighted fifty-year-old might buy a new pair of bifocals to correct his presbyopia. His glasses would contain a combination of concave, convex and cylindrical lenses to correct all of his errors of refraction.

The invention of spectacles preceded a clear understanding of the optics of the eye by several centuries. In the thirteenth century, English philosopher and scientist Roger Bacon noted that small spherical bits of glass could magnify letters and figures. He suggested that small lenses might be useful "to old persons and to those with weak sight, for they can see any letter, however small, if magnified enough." The first man to put lenses into frames and put them in front of the eyes was probably an Italian monk, Brother Alexander da Spina, who lived in the late thirteenth century. By the middle of the fourteenth century, spectacles enjoyed a surge of popularity among scholars and certain other groups. The earliest known portrait of a man with spectacles is of Cardinal Ugo di Provenza, painted by Tomaso da Modena. Some painters even adorned Biblical figures with spectacles. Venice was home to the first spectacle makers, but other manufacturers soon sprang up around Europe. For frames, craftsmen used horn, bone, leather and metal.

But since the optics of the eye were not well understood until the mid-nineteenth century, a widespread hostility to glasses persisted for centuries. Sixteenth-century oculist Georg Bartisch

106

Hermann von Helmholtz

Birth of a Science

When a ray of light strikes a human eye at just the right angle, the eye glows a dull red. The eyes of animals shine an eerie yellow. The cause of this phenomenon eluded scientists for centuries. They attributed shining eyes to phosphorescence, to electrical activity in the eye or to a reserve of sunlight stored in the retina.

Scientists who attempted to peer into glowing eyes saw only a deep velvety blackness. Yet this seeming contradiction had a simple cause. Staring directly into the eye blocked light rays that would otherwise penetrate to the back of the eye and reflect off the retina and choroid. In 1850, twenty-nine-year-old German physiologist Hermann von Helmholtz investigated this phenomenon and invented a device that opened the interior of the eye to the light of science — the ophthalmoscope.

Helmholtz was a mediocre student, probably because his classes at school did not interest him. He conducted experiments in optics at home. By the age of seventeen, Helmholtz had decided that he wanted to be a physicist. His father, however, had a more practical profession in mind, so the boy studied medicine at an institute in Berlin. After seven years as an army surgeon,

Helmholtz was appointed professor of physiology at the University of Königsberg in 1849. During his years there, he did some of his greatest work, including the development of the ophthalmoscope.

Helmholtz's original device was a crude glass and cardboard contraption. Three tiny glass plates, glued together at an angle across one open side of a black cardboard cube, reflected light from a candle into a subject's eye. Light bouncing back from the subject's retina again struck the glass plates. Though some of the reflected light glanced away, the rest penetrated the plates, a hole in the opposite side of the cube, a small lens and, finally, Helmholtz's own eye. This light carried an image of the retina and optic nerve. Helmholtz called his invention the *Augen-spiegel,* or eye-mirror.

Helmholtz's device opened a new world to him. "The appearance of the sharply demarcated red vessels upon the bright white background is of surprising elegance," he noted. He became convinced of the ophthalmoscope's usefulness as a diagnostic tool: "After one learns to recognize the characteristics of the retina in the normal eye, I have no doubt that it will be possible to diagnose all the pathologic conditions of the retina by visual observation."

Helmholtz's other achievements in optics and physics were so great that his invention of the ophthalmoscope seemed almost a sideline. Among other work, Helmholtz revived and developed Thomas Young's theory of color vision. He formulated the principle of the conservation of energy, the idea that energy can neither be created nor destroyed. Rather, it changed from one form to another. Of all the creations of Helmholtz's fertile brain, however, none has done greater service to mankind than the simple device that gave birth to ophthalmology. Since 1850, thousands of physicians have followed Helmholtz's path into the human eye and there found symptoms that enabled them to protect the treasure of human sight.

knew enough about the eye to rage against slipshod and dangerous eye operations by itinerant surgeons. Yet, he could not comprehend how poor vision might be improved by putting something in front of the eyes. "Man has two eyes," said Bartisch, "he needs not four."

The history of contact lenses is nearly as long as that of spectacles. Leonardo da Vinci, René Descartes and Thomas Young speculated the possibility of using a glass bowl or tube filled with water as an artificial lens. In one experiment, Young glued a convex glass lens to one end of a tube, filled the tube with water and held the other end over his eye. Optically, his device substituted the smooth surface of the convex lens for the irregular surface of his astigmatic eye and thus improved his vision. The same principle makes contact lenses possible.

The introduction of local anesthetics for the eye in 1884 enabled physicians to place small, curved lenses on the surface of the sensitive cornea. In 1887, Zurich physician A. E. Fick published a report of the first successful fitting of contact lenses to protect the eyes of a man with cancerous eyelids. Early contact lenses, made of glass, covered the entire front surface of the eye — both sclera and cornea. Heavy and uncomfortable, glass lenses were discarded after Kevin M. Tuohy of England patented hard, plastic lenses in 1947. Soft contact lenses followed in 1960, created by O. Wichterle of Czechoslovakia. While today's hard plastic lenses cover only the center of the cornea, soft lenses cover the whole. Containing 30 to 75 percent water, they are more flexible than hard lenses and easier to wear initially, since their flexibility permits them to conform to the shape of the front of the eye. But soft lenses are more likely to absorb bacteria and irritating chemicals from the eye or air.

All contact lenses float on a layer of mucus and tears. The liquid beneath the lens nourishes and bathes the cornea and prevents the contact lens from scratching the surface of the eye. With every blink, the eye flushes some of the fluid from beneath the lens and forces in a fresh film of tears. The tear film and lens create a new surface for the cornea and correct any optical errors of the eye. If contact lenses are worn for long periods without being removed, however, they can starve the cornea of oxygen. Bacteria and dirt can build up behind the lenses.

More than twelve million Americans wear contact lenses. But dry eyes, allergies and dusty working conditions make contacts impractical for some people; others find it difficult to insert and remove the tiny plastic disks. Many athletes, however, find contacts indispensable. And after an operation for cataracts, some patients wear contact lenses instead of thick glasses to improve the refractive power of their eyes.

Because they rest on the surface of the eye, contact lenses have potential uses that glasses cannot match. One corporation has already marketed a small plastic disk called Ocusert that can be saturated with drugs and inserted under the lower eyelid. The disk slowly dispenses the drugs and eliminates the need for eye drops. With certain improvements, regular contact lenses might serve the same purpose. Ophthalmologists and lens manufacturers are attempting to develop a contact lens that combines good optical properties with permeability to oxygen so the cornea can breathe and with resistance to the build-up of oils, proteins, chemicals and dust on the lens. If successful, they would have a contact lens that could remain in the eye for months at a time.

To See a Single Image

Refractive errors are only a few of the imperfections that can limit vision. The full definition of normal vision includes 20/20 visual acuity and the ability to use both eyes together — binocular vision. Binocular vision sends the brain two slightly different views of any object on which the eyes focus. The brain fuses information transmitted by both eyes into a single image. This fusion makes normal three-dimensional vision, also known as stereopsis, possible. If the brain cannot fuse the images from both eyes, three-dimensional vision, depth perception and the world in its full richness are lost.

At birth, a baby's eyes do not fix simultaneously on a single object. During the first year, most children develop the ability to focus on objects, hold both eyes parallel and see the world in three dimensions. But in early childhood, the

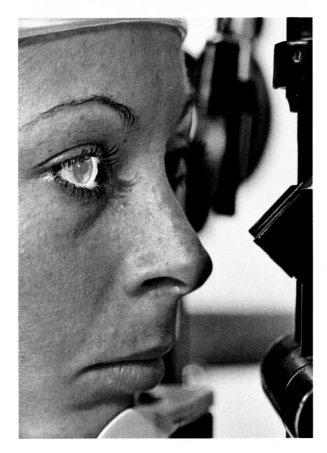

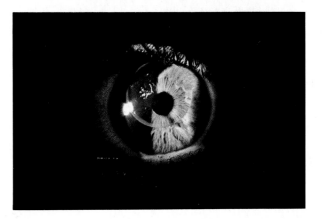

eyes of some children stop working in tandem. One eye strays slightly inward or outward or up, while the other aims straight ahead and focuses properly on objects. Occasionally, the eyes alternate roles as first one eye, then the other, strays off line. Sometimes, the lack of cooperation is intermittent, and the child sees normally as long as the eyes work together. Collectively known as strabismus, or squint, these problems affect only a small percentage of children.

A problem that often accompanies strabismus is amblyopia, a condition in which normal vision fails to develop in an eye despite the absence of disease or refractive error. For unknown reasons, the amblyopic eye sends a poor image to the brain. To protect the normal eye's clear image, the brain suppresses signals from the amblyopic eye. Since it gets no visual reward for working in tandem, the amblyopic eye may stray off line. By the same token, an eye turned inward or outward focuses on a different object than an eye aimed straight ahead. Since the brain cannot endure double vision, called diplopia, the image of the strabismic eye is often suppressed. As a result, the eye does not "learn" to see properly.

Often, doctors will patch the good eye and prescribe corrective lenses for the amblyopic or

Fitting a contact lens demands a careful check of the patient's cornea. Fluorescein dye, which glows under blue light, can outline scratches on the cornea, top left. Precision lathes used to carve contact lenses match specifications down to thousandths of a millimeter, top. Once the lens is in place, the doctor illuminates the eye with the narrow beam from a slit lamp, above, and studies the eye through a biomicroscope.

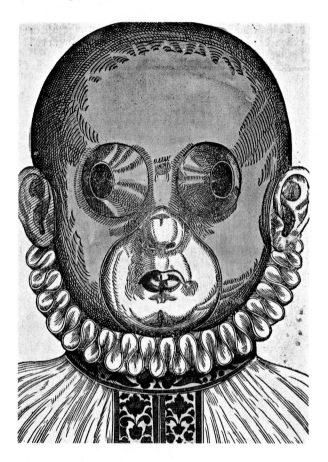

Early testing is crucial to improving vision. Children seldom complain about bad vision because they usually assume that everyone sees the same way they do. A child will not grow out of the tendency to cross his eyes and any parent who waits for the habit to disappear may be dooming a child to a lifetime of poor vision. Some doctors recommend that a child's first eye examination take place no later than age three.

Perils of Sight

If disorders of the eye stopped with refractive errors and strabismus, most people would enjoy a lifetime of nearly perfect sight. But like any other organ, the eye is subject to a variety of injuries and diseases. Injuries cause 4 percent of all blindness. An injury that breaches the cornea's outer layer might leave a permanent scar on the underlying sheet of transparent fibers, the stroma. Projectiles that pass through the cornea and injure the iris, lens or retina threaten vision itself. Of the million people who suffer eye injuries each year, 40,000 have permanent vision impairment. Yet, 90 percent of eye injuries are preventable. Careless use of contact lenses ranks second behind flying metal objects as the most common cause of injuries to the eye.

Fortunately, a scarred cornea, whether caused by injury or one of the few viruses or bacteria that attack corneal cells, is no longer a permanent impediment to vision. A cornea transplant can often restore normal sight. Donor corneas come predominantly from the eyes of people who have died within the previous day or so. With special equipment, corneas can be frozen in liquid nitrogen at −321°F and stored indefinitely. To begin a transplant operation, the surgeon extracts the donor cornea with a cylindrical cutting instrument called a trephine. Slowly rotating the trephine, which has a sharp, circular edge, the surgeon slices a round corneal patch out of the donor eye. To ensure an identical cut, the surgeon removes the clouded or scarred portion of the patient's cornea with the same trephine. He then lowers the donor cornea onto the patient's eye and stitches the patch to the surrounding tissue. When the endothelial, or inner, layer of the donor cornea is in good shape, the chances of a

strabismic eye. Eye exercises can also help correct some cases of strabismus. Sometimes, the weak eye will develop good visual acuity and allow the brain to fuse visual images. In more serious cases, complete cooperation of both eyes and normal three-dimensional vision may never be attained, no matter how long the good eye is patched.

If corrective lenses do not work, a surgeon can realign a strabismic eye by changing the lengths of the muscles surrounding it. Surgery can spare a child the emotional and psychological injury of growing up "cross-eyed" or "wall-eyed," the common names for the two most prevalent forms of strabismus, but it cannot guarantee normal three-dimensional vision. Eyes that have not learned to function together by the time the child is ten never will, even if they are aimed in the same direction. For those children who developed fusion before the onset of strabismus, however, surgery can sometimes restore normal vision.

aibue oculozum ne gru
auntur;

successful transplant are excellent. The endothelial cells of the donor's cornea regulate the flow of fluid and nutrients to the rest of the cornea and keep the tissue alive. Although the patient's own cells gradually replace the outer layers of the donor cornea, the donor's endothelium remains. Most transplant failures occur because the vital endothelial cells cannot keep the cornea alive or are rejected by the patient's own tissues. One patient underwent twenty transplant operations before regaining his sight. Approximately 75 percent of cornea transplant operations succeed.

Claiming more victims than all other eye injuries and diseases combined are the three great perils of sight: cataracts, glaucoma and diseases of the retina. Cataracts have blinded more people throughout the ages than any other affliction of the eye. Beginning as minute cloudy spots in the lens of the eye, cataracts eventually pervade the whole lens, turning it a milky yellow white, scattering incoming light and blocking vision. People with cataracts see the world as if they were looking through a waterfall, the reason ancient Greeks gave the disease its name. Research by George Benedek of the Massachusetts Institute of Technology suggests that changes in protein molecules in the lens might cause the clouding of

lens tissue. Problems that might initially trigger the formation of cataracts include too much glucose inside the lens, high doses of radiation, injuries that penetrate the lens and the long-term use of certain drugs.

Senile cataracts, so named because the slow progressive clouding of the lens usually accompanies advancing age, occur more often than any other kind. About one-tenth of one percent of children in the Western world suffer from congenital cataracts. Surgery is the only solution. More than 3,000 years ago, an itinerant Indian physician named Susruta invented an operation to move the clouded lens out of the line of sight. Using a ball-tipped knife to pierce the patient's eye, Susruta pushed the opaque lens into the vitreous cavity, out of the path of light falling on the retina. Known as couching, the operation often brought an instant improvement in vision, followed by infection and blindness a few days later. Improved versions of couching are still practiced in some countries.

The first successful operation to remove an opaque lens from the eye of a patient took place in the eighteenth century. A young painter, whose vision was clouded by cataracts, implored French surgeon Jacques Daviel to restore his

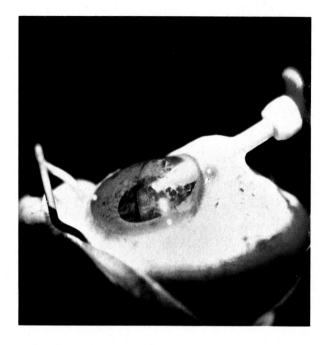 

sight. Daviel opened the eye at the edge of the cornea, cut the tiny ligaments that hold the lens in place and drew the lens from the eye. A pair of thick glasses restored the painter's sight. Daviel performed more than 400 cataract operations with only 50 failures, a record foreshadowing the high success rate of modern cataract surgery.

To remove cataracts in young patients, surgeons generally cut away the front of the lens and draw out its contents with an aspirator. A device called a phacoemulsifier, which emits high frequency sound waves tens of thousands of times a second, can help disintegrate the inner tissues of the lens so they are easier to remove. The phacoemulsifier pumps water into the lens to help liquefy the shattered tissues and simultaneously vacuums fluid from the eye. Surgeons hesitate to remove the entire lens in young patients because the vitreous usually clings to its back surface. By pulling on the front of the vitreous the surgeon might strain its connections to the retina and produce a hemorrhage or retinal tear. In older patients, however, surgeons usually remove the entire lens.

After opening the eye near the rim of the cornea, the surgeon either withdraws the lens directly or injects an enzyme, Zonulysin, that digests the threadlike ligaments holding the lens in place. Most eye surgeons now use a cryoprobe to extract the lens. The surgeon presses the cryoprobe on the lens and waits a few seconds for the tip to freeze. The lens quickly freezes to the probe. After carefully twisting the lens to break any remaining ligaments, the surgeon withdraws it from the eye. Since the iris and the vitreous sometimes press forward with the removal of the lens, surgeons cut away a small piece of the iris, a procedure called an iridectomy. An iridectomy ensures that fluid in the eye circulates properly.

With the removal of a cataract, the optics of the eye change forever. In the past, only glasses could restore a patient's sight. But the thick lenses magnify the patient's world by a third, play havoc with depth perception and distort peripheral vision. The introduction of contact lenses offered improved sight for many cataract patients since contacts provide a broader visual field and much less distortion. Since the 1950s, surgeons have been implanting intraocular lenses in the eyes of some patients during the cataract operation. Tiny disks of plastic, the intraocular lenses either clip to the iris, almost like a paper clip on a page, or rest on two small legs that extend behind the iris into the vacant lens capsule.

112

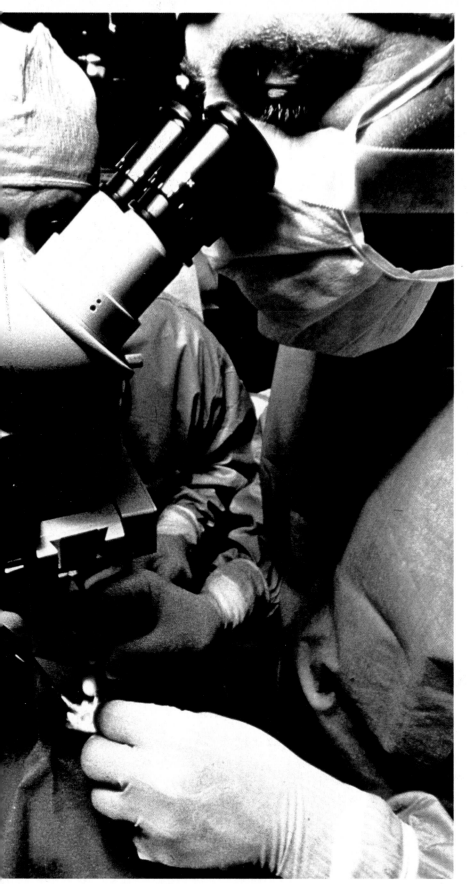

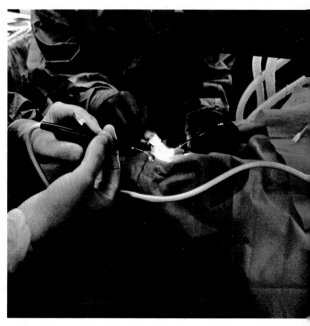

Operating microscopes permit surgeons to repair delicate tissues with surgical thread finer than a human hair. Microsurgery is especially valuable in healing the eye. Ophthalmologists routinely employ operating microscopes in cataract operations, left. After the clouded lens has been gently loosened from its moorings inside the eye, many surgeons use a cryoprobe to extract the lens. As it touches the lens, the tip of the cryoprobe is supercooled, and the lens freezes to the probe. Carefully withdrawing the cryoprobè, above, a surgeon removes the diseased lens from the eye.

Of the nearly 500,000 cataract operations performed in the U. S. every year, about 95 percent are successful.

The most common form of glaucoma steals sight gradually, and has thus earned the grim nickname "sneak thief of sight." At least two million Americans suffer from the disease. If glaucoma is diagnosed early, drugs can control it for a lifetime. People with undetected glaucoma can lose much of their vision before realizing how severely the disease has restricted their sight. Consequently, physicians recommend that everyone over thirty-five have his eyes tested for glaucoma at least every two years.

The major sign of glaucoma is high pressure within the eye. The rise in pressure results from a build-up of aqueous fluid. This fluid bears a heavy responsibility. The nutrients it contains feed both the cornea and the lens. The ciliary body, behind the iris, constantly secretes aqueous fluid, about one-fifth of an ounce a day. From the ciliary body, the fluid flows into the posterior chamber, then slowly circulates over the lens and toward the pupil. There, it flows over the rim of the iris and into the anterior chamber, behind the cornea. At the outer edge of the anterior chamber, where the iris meets the back of the cornea, lies the trabecular meshwork, a webbing of tiny fibers and canals that steadily drains the aqueous fluid out of the eye. If these drainage canals are blocked, pressure rises and squeezes the tiny capillaries that feed the blanket of microscopic nerve fibers within the eye. With the passage of time, some of the nerve fibers, usually those responsible for transmitting peripheral vision, die. The final stages of glaucoma are acute tunnel vision and sometimes blindness.

Some eyes tolerate high intraocular pressure better than others. In fact, there is no perfect standard for eye pressure. Sometimes, glaucoma first becomes apparent by the damage it causes, such as a slight decrease of peripheral vision. Another sign is a change in the shape of the optic nerve, visible through the ophthalmoscope as a pale disk on the retina.

Glaucoma takes two forms. Open-angle, or chronic, glaucoma accounts for 90 percent of all cases. Angle-closure, or acute, glaucoma is an ex-

plosive and much rarer form of the disease. It can cause blindness in twenty-four to forty-eight hours if not treated. Intense pain and nausea accompany acute glaucoma, as well as intraocular pressure so high that the eyeball feels as heavy as a stone. Acute glaucoma often develops when fluid pressure or a spasm pushes the iris into the "angle" of the eye, the spot where the iris meets the cornea. There, the iris completely blocks the eye's drainage system. Physicians usually administer miotic drugs, such as pilocarpine, to constrict the iris and pull its base out of the angle of the eye. Usually an iridectomy follows. The surgeon opens a tiny hole in the iris so that aqueous fluid can flow easily from the posterior chamber to the trabecular meshwork and out of the eye.

Open-angle, or chronic, glaucoma presents a more difficult problem, since its effects appear only gradually, insidiously. Science has yet to discover why the drainage systems of some eyes fail. The most common treatment for chronic glaucoma is a regimen of drugs — pilocarpine or other miotics to constrict the pupil, epinephrine or acetazolamide to slow the secretion of aqueous fluid or a combination of other drugs. A small pharmacological battle rages today over the merits of marijuana as an antiglaucoma drug. Tests with its active ingredient, THC, have failed to reveal any benefits, but smoking marijuana regularly seems to lower eye pressure in some people. A new drug, Timolol, inhibits the secretion of aqueous fluid without the side effects that sometimes accompany the use of other drugs.

If drug therapies fail, surgery for open-angle glaucoma can often help. One method is to cut a tiny hole in the eye, just outside the rim of the cornea. The conjunctiva absorbs aqueous fluid as it drains through the hole. A more drastic procedure involves removing some of the trabecular meshwork or cutting through to the canal of Schlemm, a deeper part of the eye's drainage system. In the last decade, a few eye specialists have begun using lasers to burn tiny holes in the webbing of the trabecular meshwork. Any of these operations can slow the progress of glaucoma, but drugs remain the treatment of choice.

More than the cornea or the lens, remarkable structures though they are, the retina is the eye's

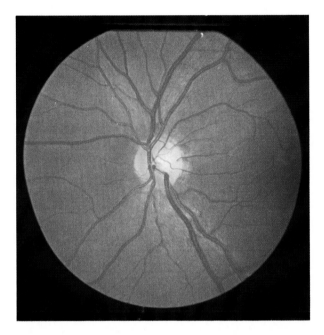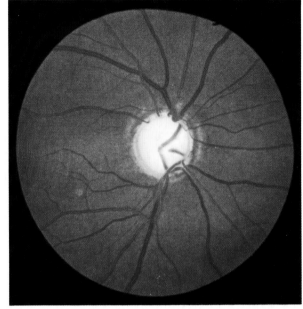

miracle. No other tissue in the body demands as much oxygen or food to maintain its labors. And the retina's complexity is the complexity of the brain. Indeed, the retina is part of the brain, turned outward to face the world. The retina's intricacy makes understanding, protecting and healing it a huge task. Hypertension, blood clots and other afflictions can all damage its fragile fabric. The retina also suffers from diseases that appear to have their origins in its microscopic depths. Retinitis pigmentosa, a malfunctioning of the rods and cones, is a disease that doctors still do not fully understand. The disease strikes few people, but usually results in blindness. Retinitis pigmentosa attacks the midperiphery of the eye first, carving a circular band of blindness in the visual field. Peripheral vision degenerates next, and finally central vision fades.

Retinal detachment is a side effect of some retinal diseases and an occasional result of a violent blow to the eye. The vitreous sometimes tugs on the retina at one of the sites where the two meet and rips a small, horseshoe-shaped tear in the retina's delicate fabric. If vitreous fluid seeps behind the retina and forces it away from its underlying layer, the pigment epithelium, a retinal tear opens into a retinal detachment. Through an eye

In a healthy eye, left, the blood vessels that supply the retina rise out of the pale optic disk at its center. The optic disk is the passageway into the eye for retinal nerve fibers and blood vessels. In an eye stricken with glaucoma, the high pressure of fluids inside the eye presses on the delicate nerve fibers and partially excavates the optic disk. The elevated intraocular pressure also forces retinal blood vessels to climb up the side of the hollowed-out optic disk and curl over its rim as they enter the eye, right. Cupping of the optic disk, as this symptom is known, is a cardinal sign of glaucoma.

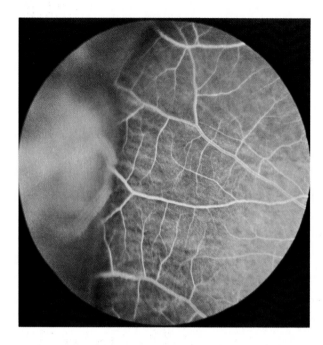

with a detached retina, the world seems half-dark and half-light, as if someone had thrown a veil over half of the visual field of that eye.

To repair a detached retina, surgeons use laser beams, cryoprobes or intense heat to destroy tissue beneath the tear and rejoin the layers. In cases of severe retinal detachment, the surgeon must use a small silicone implant to bring the outer layers of the eye close enough to the detached retina for the layers to heal together. About 85 percent of all surgery to heal detached retinas succeeds.

Although many diseases threaten the retina, two afflictions cause more blindness than any other disorder of the eye — diabetic retinopathy and senile macular degeneration. Diabetic retinopathy blinds more people between the ages of twenty and seventy-four than any other disease.

Diabetes attacks the vital capillary network of the retina. Diabetic retinopathy takes two forms, proliferative and nonproliferative. Many of the same symptoms occur in both types of the disease. In addition to the destruction of capillaries, tiny white spots appear on the retina where fats or fluids leak out of the damaged blood vessels. Other symptoms include patches of damaged nerve tissue called cotton-wool infarcts because

of their woolly, white appearance and microscopic hemorrhages and microaneurysms that leak blood into the retina. In people whose diabetes begins in adulthood, the disease often stops with these symptoms. When this occurs good vision may remain. In some cases of nonproliferative retinopathy, surgeons use lasers to destroy patches of tissue on parts of the retina. By sealing off some blood vessels to reduce the retina's demand for nutrients and oxygen, they hope to halt the progress of the disease.

If diabetic retinopathy progresses to its proliferative stage, however, the outlook is worse. In some people who have suffered from diabetes since childhood, the insufficient blood supply to the retina apparently creates an insatiable demand for food and oxygen. New blood vessels supported by a net of fibrous tissue proliferate across the retina and the back of the vitreous. Injected into the blood stream, fluorescein dye clearly reveals the progress of the disease by outlining new retinal blood vessels and pinpointing tiny hemorrhages. As its structure weakens with age, the vitreous may pull on these new blood vessels, causing them to rupture and bleed into the vitreous cavity. Diabetic retinopathy sometimes blinds by obscuring vision with a dark cloud of vitreous fluid and blood. Photocoagulation with a laser can slow the advance of the disease and a vitrectomy often restores some useful vision. But proliferative diabetic retinopathy is frequently an irreversible affliction. The cure for this eye disorder awaits the cure for diabetes.

Old age takes a toll on the retina, as it does on all living tissue. In many people, the hardest-working region of the retina, the macula, gradually degenerates with age. Most of the macula's nourishment comes from the blood vessels of the choroid by way of the pigment epithelium layer. Blood flowing through the choroid also absorbs heat from the macular cells. Since it lies directly in the path of light rays focused by the eye, the macula runs a greater risk of suffering heat damage than any other part of the retina. Thickening of the choroidal blood vessels can lead to a breakdown in the transfer of nutrients and oxygen from choroid to macula and can slow the equally vital absorption of heat.

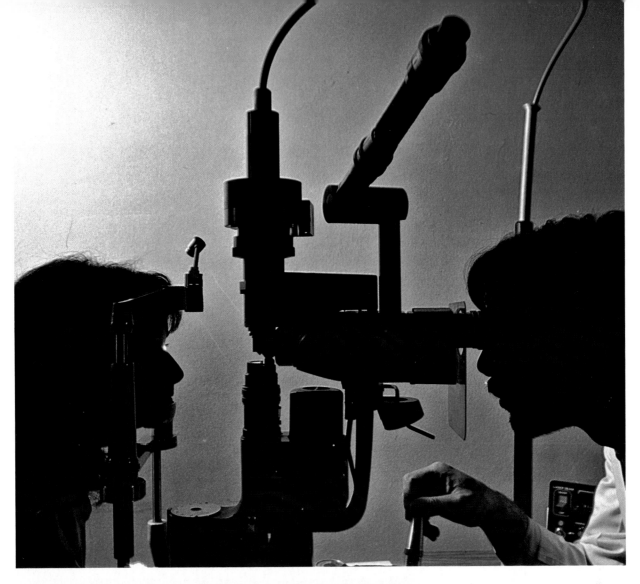

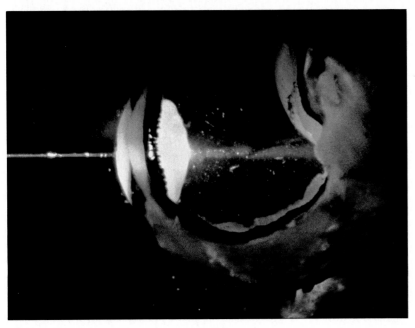

In the hands of a physician, the laser can be a healing light. Its intense beam can safely breach the cornea and lens and touch the innermost tissues of the eye, left. Some cases of retinal detachment, diabetic retinopathy and glaucoma yield to laser treatment. Simple and painless, the procedure can often be performed in a few minutes at an ophthalmologist's office.

Arteriosclerosis, hypertension and old age can all cut down the macula's vital supplies. Beneath the degenerating macula, small white patches known as drusen become visible in the ophthalmoscope. This condition eventually leads to a slow, steady decline in central vision. Some degree of macular degeneration afflicts almost a third of all Americans over sixty-five.

Retinal disease, glaucoma and cataracts, in that order, are the leading causes of blindness in the United States. Yet, blindness resulting from glaucoma and cataracts can often be prevented. Diagnosis and treatment of these diseases could have preserved good vision in many of the 11.4 million Americans classified as visually impaired. More than a million of these people have a severe impairment, leaving them unable to read a newspaper even with glasses or contact lenses. About 500,000 Americans are legally blind. They have less than 20/200 vision in the better eye or a severely restricted field of view, a condition that might be described as acute tunnel vision.

Resources for the Blind

The loss of sight inevitably involves losses of other kinds believes Father Thomas Carroll, an expert on blindness and rehabilitation of the blind. Security, mobility, trust in the other senses, a feeling of physical integrity, independence and the ability to communicate with the written word disappear when sight disappears. The notion that blind people gain more acute senses of smell, taste, touch and hearing is false. Blind people may pay closer attention than those with sight to information from their other senses, but their hearing, smell and taste are no more acute than anyone else's. In a disease such as diabetes, the effects on small blood vessels throughout the body can damage all the senses. Not only are gains from blindness accidental, says Carroll, "they are by no means universal, . . . where they exist they are occasioned by blindness rather than caused by it."

The loss of sight, however, is not the loss of perception. People who lose their sight retain a mental image of the world and still receive countless sensations from other senses. The world inside the brain, the perceptions built from

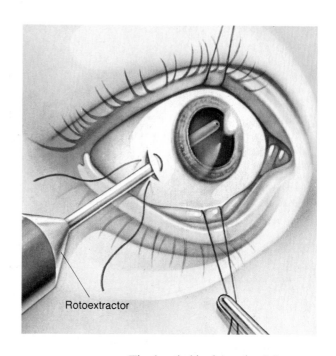

Rotoextractor

The fragile blood vessels of the retina sometimes leak blood into the clear, jellylike fluid of the vitreous as a result of diabetic retinopathy and other serious eye problems. Too much blood in the vitreous clouds the fluid, blinding the eye. When the eye's normal ability to absorb blood from the vitreous cannot clear vision, surgeons sometimes attempt a vitrectomy, a desperate surgical effort to save sight. To draw out the vitreous, surgeons insert a small cutting tool into the eye, above. A rotating edge in the center of the tip snips off globules of vitreous, and a suction tube draws the dark jelly from the eye. As the tool removes the clouded vitreous, it constantly refills the cavity with a saline solution to preserve the eye's shape. Because the vitreous attaches in some places to the retina, surgeons attempting a vitrectomy must take great care to keep from damaging the retina's delicate tissues or rupturing still more blood vessels, right.

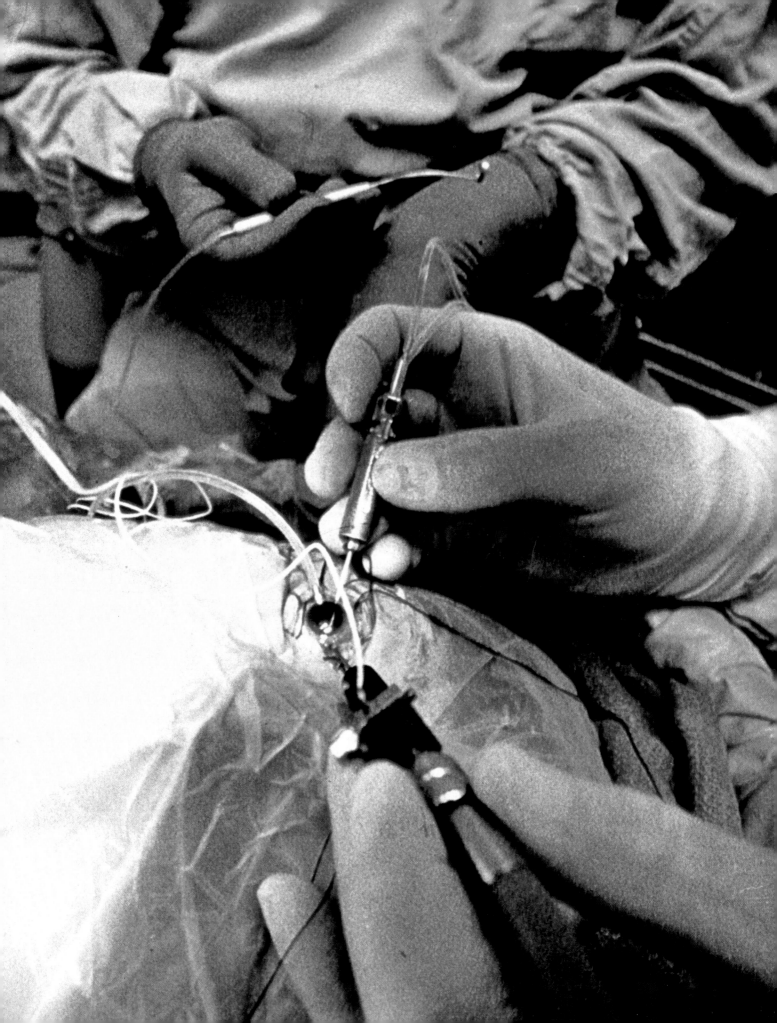

With the help of electronics, a small camera mounted on eyeglasses and an array of metal stimulators touching the torso, the skin can "see" when the eyes fail. Biophysicist Carter Collins and colleagues have developed a series of experimental visual aids for the blind. In this early version, microcomputers on the vest and the stimulator array convert the camera's images into tactile drawings on the skin.

a lifetime of sensation, can be a resource to the blind. Scientists are seeking to tap this inner world by building devices to aid the blind. Some experimental devices can read a book and instantly transcribe it into braille. Others tap out the letters of a book on a person's fingertips. Scientists are also working on a device to reproduce entire written sentences in a computerized version of human speech. Sonar spectacles that bounce sound waves off surrounding objects might someday help blind people navigate.

Californian biophysicist Carter Collins has developed a Tactile Sensory Replacement (TSR) system to substitute for sight. A small television camera mounted on a pair of glasses picks up images that are translated by a small computer into a pattern of electrical charges. The computer then triggers any of a thousand metal stimulators strapped to the subject's torso. If the camera sees a triangle, the metal stimulators set up a pattern of vibrations in the shape of a triangle on the subject's skin. Both the congenitally blind and people who have lost their sight could use the device to identify common shapes and objects. Some of Collins's subjects have developed enough expertise with the machine to assemble tiny microcircuits.

Collins's most recent creation, developed with the help of computer scientist Michael Deering, translates vision into both sound and touch. The device consists of a shoulder-mounted camera, a portable computer and sixteen metal stimulators touching the patient's skin. If the camera detects a telephone pole in the blind person's path, it transmits the image to the portable computer, which converts the image into the word "pole" in computer speech. At the same time, the computer sends a message to the metal stimulators, which tap out the direction and distance of the pole. Both of Collins's devices are still experimental, but they may someday provide a new kind of "sight" for the blind.

For the most part, however, modern ophthalmology aims not at bypassing the eye but at protecting and healing it and improving its sight. Ultrasound, the use of high frequency sound waves to "see" the soft tissues of the body, has proven especially useful for peering into the eye.

A small transducer beams sound waves into the eye and records their reflections as they bounce off the cornea, lens, retina and other tissues. From the recordings, scientists can produce a cross-sectioned image of the eye. Ultrasound can even produce a portrait of an eye so clouded with cataracts by a vitreous hemorrhage that no ophthalmoscope could penetrate its opacity. Tumors of the choroid or the orbit of the eye often show up clearly on ultrasound scans. Ultrasound can pinpoint metal chips or other small objects that have penetrated both cornea and lens and disappeared into the vitreous. Retinal detachments and dislocated lenses will also appear on an ultrasound scan. Outside the eye, ultrasound fashions images of the optic nerve, muscles, orbital fat and other structures.

Carving the Cornea

Ophthalmological surgery has made tremendous strides during the last few decades. A new substance called Healon, a purified extract from the combs of roosters, is improving the success rate of eye surgery. One risk in eye operations is loss of fluid from the anterior chamber, which might cause the vitreous, iris or other eye structures to move out of position. Healon, developed by Endré Balazs of Columbia University's Department of Ophthalmology, helps gently push the vitreous, the iris, even the retina back into position and hold them in place. Healon may be most useful in cataract operations. Injected into the anterior chamber, Healon prevents the lens or vitreous from squeezing forward and chafing the fragile, indispensable cells of the corneal endothelium. Lens implants stay in better position with the use of Healon. By preventing dislocation of the lens and iris, Healon helps keep drainage canals open and reduces the risk of postsurgical glaucoma.

More controversial developments in eye surgery are designed to improve the vision of nearsighted or farsighted people. An operation called keratophakia changes the shape of the cornea to improve poor vision. In keratophakia, developed by José Barraquer of Bogotá, Colombia, the surgeon slices off a small portion of the patient's cornea, slips a disk from a donor cornea underneath and then stitches the sliced section back on.

The optacon transforms letters on a printed page into sensations on the fingertip. Holding a small camera in one hand, a blind reader slowly scans a page of text. Inside the optacon, 144 miniature rods impress the outlines of the letters of the alphabet — one at a time in the order they appear on the page — onto one of the fingertips of the reader's other hand.

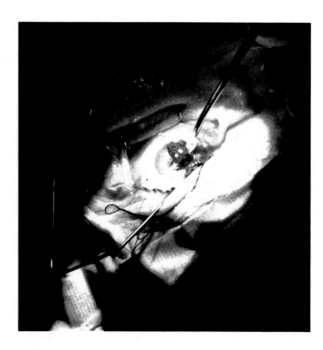

A carefully carved chip from a donor cornea, tinted blue to make the transparent tissue more visible, is lowered onto a patient's cornea, where it helps reshape the front surface of the eye and increase the eye's refractive power. After a cataract operation, the lensless eye loses some of its refractive power and becomes extremely farsighted. People who cannot tolerate intraocular lenses, contact lenses or thick heavy cataract spectacles may have a fourth alternative in this new eye operation — keratophakia. By shaving off a slice of the cornea, inserting a section of a donor cornea and stitching the slice back in place, surgeons can change the cornea's shape and restore the eye's ability to focus light.

The donor cornea has been carved on a computerized lathe to increase the curvature of the patient's cornea and give it more refractive power, improving farsightedness. In another of Barraquer's operations, keratomileusis, the surgeon slices off a piece of a patient's cornea, freezes it and reshapes it on a lathe. The section is then sewn back onto the eye. Keratomileusis can be used to improve either farsightedness or nearsightedness.

Barraquer's operations have not engendered the storm of controversy that currently surrounds an eye operation created by Soviet ophthalmologist Svyatoslav Fyodorov. In 1973, a sixteen-year-old student named Boris Petrov visited Fyodorov. Petrov's glasses had been shattered in a fist fight and shards of glass were embedded in his cornea. Fyodorov removed the glass and the eye healed, but remarkably, the boy's myopia also disappeared. Fyodorov thought that "if a boy can treat myopia with his fist, maybe we can treat it surgically." His operation consists of making eight to sixteen cuts in the cornea. The cuts radiate like the spokes of a wheel from a central hub of untouched tissue. The incisions are of different lengths and depths depending on the degree of the patient's myopia. After the operation, internal eye pressure flattens the central portion of the cornea and reduces the degree of myopia. More than 4,000 radial keratotomies, or radial k's, have been performed in the U. S. since 1978. Fyodorov and his colleagues have performed another 4,000 since their first attempt in 1975. In more than half of the cases, the patients do not need glasses after the operation. For most of the others, vision improves to some degree, although some still must wear glasses. Radial k's require only local anesthesia, take about forty-five minutes in the ophthalmologist's office and cost about $1,000 an eye.

For people whose occupations require good vision without glasses, radial k's could prove valuable. Many ophthalmologists, however, condemn radial k's as "buccaneer surgery." They decry the the lack of long-term studies of the surgery's effect on the eye. While some patients are satisfied with the operation, others have complained of glare and fluctuating visual acuity throughout the course of the day. Another criticism of radial keratotomy is that it is unnecessary. One surgeon trained in the technique admits the operation probably works, but says he hasn't found anyone who really needs it. The National Eye Institute has begun a $2.4 million study at eight major hospitals over the next five years to test the operation's safety and effectiveness.

How great a role do visual disorders play in life? Blindness is a loss that can only be endured. Given the neurological fact that half the fibers bringing sensation to the brain stem from the optic nerves, nearsightedness, farsightedness and astigmatism may all affect the way we perceive the world. In *The World Through Blunted Sight*, British ophthalmologist Patrick Trevor-Roper writes, "The myopic eye, initially but one facet of an inherited mould of the human frame, may, in a limited way, continue to influence the development of that frame, its posture and its movements throughout life. More especially is the myopic eyeball a part of the personality structure, and its influence on the evolution of this personality may be of paramount importance."

The imprint of the flawed eye appears in the works of poets, painters and other artists. John Keats, allegedly nearsighted, wrote more often of objects near at hand, as in his ode to a Grecian urn, than of far-off vistas. The dark browns and somber blacks of James Whistler's paintings may reflect both a brooding personality and his suspected colorblindness. Trevor-Roper believes that Cézanne, Renoir, Degas and Pissarro were all nearsighted. "The high frequency of myopes among artists, and among the Impressionists in particular," he says, "is probably not just coincidental." The canvas may well have been clear to the myopic eye, although the subject was not. Trevor-Roper suggests that the Impressionists' emphasis of shade and outline over line and form may have been the result of both aesthetics and optics, the mind's eye and the eye itself. Trevor-Roper does not push his conclusions too far. Personality and profession are not solely the products of good or bad vision. But sight is an integral part of life, the preeminent sense. What affects the eye will surely affect the person. Or in the words of English poet William Blake, "The eye altering alters all."

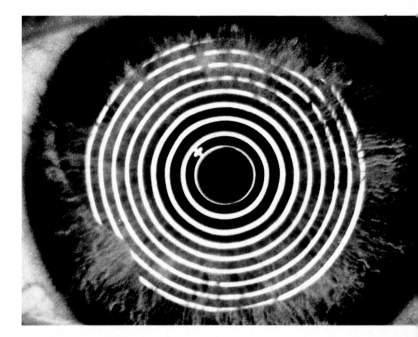

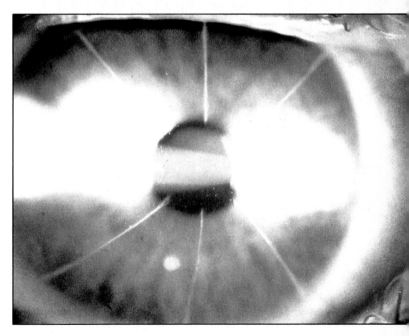

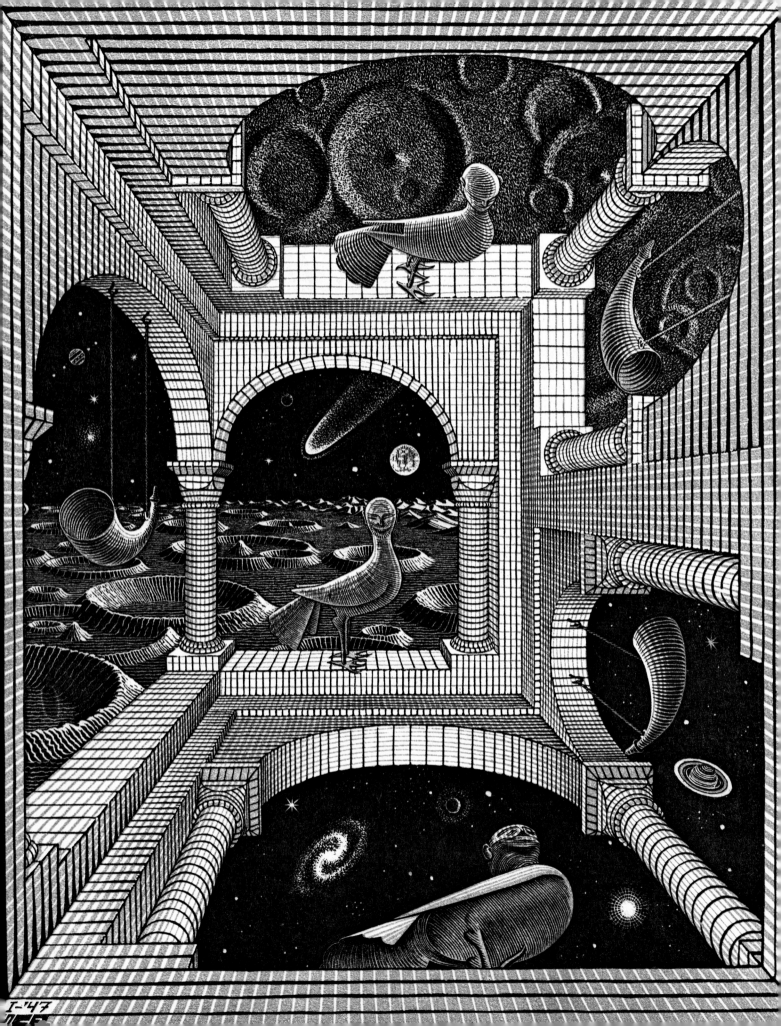

I-'47

Chapter 6

Illusion and Artifice

In 1906, Arctic explorer Robert E. Peary stood on a summit at the glacial fringe of Cape Thomas Hubbard and surveyed the ragged surface of the polar pack. Holding field glasses to his eyes with clumsy fur gloves, Peary surveyed the largest unexplored expanse of ice on earth. Then, over the polar sea, 120 miles to the northwest, he sighted a region of extraordinary beauty above the ice horizon. The snowcapped mountains appeared to reach an altitude of more than 15,000 feet, more than twice the height of the tallest peaks encountered anywhere in that region. "My heart leaped the intervening miles of ice as I looked longingly at this land," he wrote, "and in fancy I trod its shores and climbed its summits, even though I knew that the pleasure could be only for another in another season." Peary named the formation of icy peaks Crocker Land, honoring one of his financial backers. With a final wistful look, the fifty-year-old explorer trudged on. Three years later, Peary finally reached the North Pole, but he left the search for Crocker Land to a younger man.

In 1913, having obtained the support of several museums, including the American Museum of Natural History, Donald B. MacMillan set out with 6 expeditionary colleagues, 15 Eskimos and 150 dogs to find and map Crocker Land. The explorers headed toward the northwest, over hard, rolling surfaces of blue ice. Finally, with 500 miles and weeks of bad weather behind them, they reached the point where Peary had sighted land that some geologists declared could not exist. Mountain ranges of such magnitude, skeptical scientists said, could not stand in the Arctic.

On the first sunny day, the cry, "We have it!" rang out. "We ran to the top of the highest mound," MacMillan wrote. "There could be no doubt about it. Great heavens! What a land! Hills, valleys, snowcapped peaks extending through at least 120 degrees of the horizon." He

Dutch artist M.C. Escher's Other World *draws us into a universe that distorts visual reality. The conflicting vistas make it impossible to decide whether we are perched on a ledge looking down or up.*

125

The apparently immense horizon moon rises over Monument Valley, Arizona, below. The photograph vividly captures the effect of the moon illusion — a riddle that has puzzled man since antiquity.

turned to an Eskimo and anxiously asked which course was best, but the man only said, *"Poo-jok,"* meaning mist. MacMillan, astounded by this reply, nevertheless ordered the expedition forward. As the men proceeded, the landscape gradually changed. The majestic mountains faded before their astonished eyes until finally, as the sun moved toward the evening sky, Crocker Land disappeared altogether. The explorers plodded on but found only a barren wasteland of broken ice. His hopes dashed, MacMillan confided in his journal, "You can imagine how earnestly we scanned every foot of that horizon — not a thing in sight, not even our almost constant traveling companion, the mirage. We were convinced that we were in pursuit of a will-o'-the-wisp, ever receding, ever changing, ever beckoning."

Both Peary and MacMillan had witnessed what has come to be known as Fata Morgana, a spectacular mirage named for the nymph who, in Arthurian legends, created castles in the air. The explorers had observed Crocker Land through a portion of the atmosphere that acted as a giant lens, bending the light rays that passed through it. The image they saw was real, but it was not cast by true peaks and valleys. The illusory "snow-clad summits" were the inverted reflections of clouds and ice.

The Moon's Ancient Riddle

Visual illusions, whether natural or contrived, have deceived and fascinated man for centuries. Perhaps the most familiar is the moon illusion. When looming over the horizon, the moon appears larger than when it is high in the heavens. This phenomenon occurs even though the image the moon casts in photographs and on the retina of the eye remains unchanged. To solve the moon illusion, psychologist Edwin G. Boring and his colleagues at Harvard University conducted a

126

series of experiments. The researchers asked subjects to look straight ahead at the horizon moon and then to gaze upward when the moon reached its zenith, the highest point in the sky. Most observers thought the moon they saw, when looking up, appeared one-and-a-half to two times smaller than the moon on the horizon. Boring then asked two people to lie down so that they could view the zenith moon by looking straight ahead. To see the horizon moon behind them, the observers tilted their heads backward and raised their eyes. They reported that the horizon moon appeared smaller — the illusion had reversed. Based on these and other tests, Boring concluded that the elevation of a person's eyes affected the perceived size of the moon.

Intrigued by Boring's theory, Lloyd Kaufman of Yeshiva University and Irvin Rock of the New School for Social Research attempted the same experiment. The psychologists observed the moon while lying down but failed to achieve Boring's results. Kaufman and Rock then fashioned a device which projected an image of the moon against a partially mirrored glass plate. One artificial moon could be aimed at the horizon and another at the sky's zenith. Each device had a set of variable apertures that enabled the observer to adjust the sizes of the moons until they matched. Observers first gazed straight ahead at the artificial horizon moon and compared it with a zenith moon which they viewed with elevated eyes and, then, with level eyes. They experienced the illusion both ways. Next, viewers compared two moons in the same region of the sky, first looking straight ahead, then with eyes raised. The observers noticed only a slight difference in size, not enough to account for the dramatic illusion normally seen.

The scientists turned to a theory proposed by Ptolemy nearly 2,000 years ago. The Greek as-

127

tronomer had claimed that the moon seen at the horizon, across fields, looked more distant than when viewed against the perceived flatness of the zenith sky. Since, in reality, the moon's size remains constant, the image appearing more distant will seem larger. Yet many observers argue that the "larger" horizon moon seems closer, not farther away. The psychologists asked ten people to scan a moonless sky and imagine it as the inside of a dome. They were asked whether the surface seemed closer at the horizon or at the zenith. Nine of the observers thought the horizon sky seemed farther away. If the upper region of the sky is seen as a flattened vault with the moon affixed to its surface, the moon appears more distant at the horizon, and therefore larger, whether the observer realizes it or not.

Pointing both artificial moon devices at the horizon, observers viewed one lunar image through a hole in a sheet of cardboard that isolated the moon by masking the terrain. The other device presented the moon over the city's skyline. Observers matched the two disks by enlarging the masked image to a size that was actually much larger than the other. When both moons were isolated, no illusion occurred. Yet, the horizon moon does not look larger solely because it can be compared with structures or trees. The illusion also occurs over water or desert where there are no familiar objects for comparison. Ptolemy's theory stresses the impression of distance created by terrain. If apparent distance were the source of the illusion, Kaufman and Rock reasoned, then viewing the horizon moon upside-down should decrease the illusion, since image inversion decreases one's sense of distance. The scientists took a group of people to a New York City rooftop where the horizon moon could be seen framed between tall buildings. Bending down, the observers looked backward between their legs and reported a reduced illusion. Kaufman and Rock's experiments show that Ptolemy's theory was based on sound scientific reasoning. Other scientists, however, remain unconvinced that the "apparent distance" theory explains the ancient lunar enigma.

Ptolemy and others might have puzzled over illusions in ancient times, but systematic investigation of visual distortions began only a century ago. The invention of popular optical devices like the kaleidoscope and the stereoscope, which imparts a three-dimensional effect to photographs, fostered public curiosity about the perceptual "magic" of illusions. Some researchers suggest that the invention of improved writing tools in the 1850s led to an increase in doodling and the chance discovery of many geometric illusions. At about the same time, students of the newly founded science of psychology began analyzing visual illusions. Between 1860 and 1900, several hundred scientific papers discussing geometric illusions were published. In 1915, a radical new school of psychology was founded. Now known as Gestalt psychologists, they departed from established scientific convention. Believing that "the whole is different from the sum of its parts," Gestalt psychologists studied total visual experience instead of first considering separate visual elements. They considered illusions to be actual but erroneous perceptions and developed theories which have since been used by other psychologists to explain illusory perceptions.

Since 1950, scientific interest in visual illusions has risen steadily. As neurophysiologists began to probe the neurons in the visual pathway, psychologists and physiologists pooled their efforts. Both disciplines investigate vision and the ways in which illusions trick the eye, for such discoveries provide important clues to man's perception of the world. Scientific efforts, though, have yielded few answers that satisfy everyone.

The Mind's Best Bet

Fraser's spiral remains as puzzling as in 1908, when British psychologist James Fraser presented the oddity to his colleagues. Careful tracing of the circles proves the spiral exists only in the viewer's mind. Many experts suggest explanations that echo Gestalt doctrine. According to Gestalt theory, the mind perceives lines of similar direction in a coherent or unified group. Modern theorists speculate that the visual system mentally collects the inward sloping line components of all the concentric circles into one continuous line. The instant the mind reaches this conclusion, it succumbs to the illusory spiraling effect provided

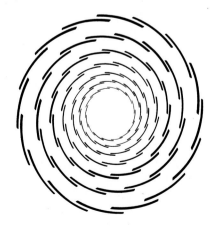

Careful tracing reveals that Fraser's spiral is composed of concentric circles. But the spiral effect is so powerful it can even trick the finger. Without the checkerboard background, below, the illusion vanishes.

by the directional cues in the receding checkerboard background. Still, this does not explain why the visual system takes such elaborate steps instead of seeing the concentric circles that are there. Nor does it adequately explain why the illusion continues to fool the visual system even after the observer realizes the spiral does not exist. Some scientists believe the mind continually forms hypotheses about reality based on visual signals combined with its vast store of past experiences to choose what psychologist Richard Gregory calls its "best bet." Fraser's spiral apparently generates an unshakable, yet inappropriate "theory" of reality. When it deceives the mind's thought processes, an illusion is cognitive.

Some images conflict too intensely with the mind's theories of reality. When a person looks at the mold of a human face, his mind rejects what his eyes see. Since no such face exists in our three-dimensional world, the mind "sees" a normally protruding brow, cheeks, nose and chin, even though the sense of touch can aid in telling the brain otherwise.

Sometimes the visual system forms two hypotheses of reality for the same situation, each one equally plausible. A piece of paper folded lengthwise down its middle and placed before an observer looks like a tent; the mind develops a hypothesis that this is so. But if the observer closes one eye and stares at a point along the paper's midline, the mind forms a hypothesis that the object is the inside of a corner. The image cast on the retina by the tent is the same mental picture a corner presents. Since both hypotheses are correct, the mind wanders between the two choices. It selects first one image and then the other, continually testing their veracity.

In the late forties, visual researcher Adelbert Ames, Jr., investigated the concept of mental hypotheses. An ophthalmologist, Ames treated patients suffering from aniseikonia, a disease that distorts vision by creating images of different sizes in each eye. Fascinated by the effect of the disease on depth perception, Ames designed a series of visual illusions to demonstrate how the mind forms hypotheses about reality and depth. In one display, an observer peers through a peephole and sees a chair in the corner of a room. But

when he views the scene from above, he realizes that the chair is composed of segmented forms suspended by thin wires. Only from one view do the lines of the objects converge to form the shape of a chair on the eye's retina.

Another of Ames's illusions demonstrates how contrast in size leads to mental hypotheses about distance. Two balloons of equal size, color and illumination are placed in a darkened room. As one balloon is inflated and the other, deflated, the balloons appear alternately to approach and retreat. The illusion exploits the mind's hypothesis of "size constancy." The perceived size of an object remains constant even though the image grows as the object approaches and shrinks as it retreats. Were the mind not equipped with such a hypothesis of reality, a man walking toward an observer would appear to increase in size.

The mind also judges depth by considering brightness, overlap and parallax — the apparent change in an object's location created by a change in the observational angle. Ames tested these aspects of depth perception in a specially designed, distorted room. He constructed a trapezoidal room for the observer to view by gazing through a peephole. In reality, one corner of the room is farther from the observer than the other corner. But when viewed from a particular angle, the room appears normal. If a man stands in the far corner and a small boy in the near corner, the boy appears larger than the man because of the distorted perspective. Although the observer knows that adults are usually bigger than children, he still experiences the illusion.

Tricking the Eye's Hardware

The Ames demonstrations test the mind's interpretation of visual signals. Other illusions deceive mechanisms of the eye. When an illusion tricks the eye's "hardware," it is physiological in

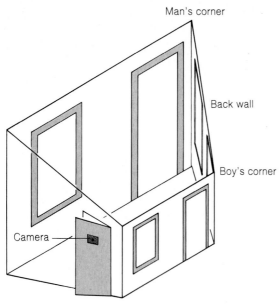

Man's corner

Back wall

Boy's corner

Camera

The children at far right appear bigger than the adult on the other side of the Ames distorted room. Actually, the man is taller than any of the children. The diagram shows the actual layout of the room.

origin. The lens casts an inverted image of the prevailing scene onto the light-sensitive retina at the back of the eye. The retina contains more than 100 million photoreceptors, which absorb light and fire impulses in response to it. These signals travel from photoreceptors to bipolar cells and then to the ganglion cells that form the optic nerve. Each ganglion cell possesses a receptive field composed of clusters of photoreceptors. Light striking the center of a receptive field may spark impulses that course along the optic nerve to the visual center in the brain to indicate the presence of bright areas. Light hitting the edge of such a receptive field prevents the ganglion cell from firing. If enough light strikes both areas of the receptive field, the antagonistic responses suppress the ganglion's impulses, sometimes stopping the neuron from firing altogether. The visual cortex in the brain interprets the signals in terms of contrasts between light and dark areas. This process, called lateral inhibition, gives rise to a variety of visual illusions.

The Hermann-Hering grid probably depends on such interaction among nerve cells in the retina. A white grid placed against a black background, the drawing induces the appearance of illusory gray spots at the intersections of the

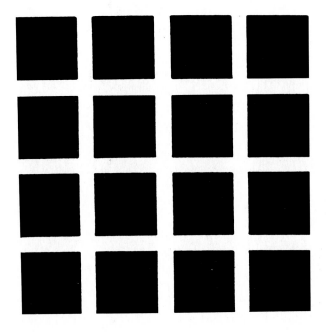

white lines. When the upside-down image of the grid becomes focused on the retina, the light reflecting from the intersecting white lines strikes the inhibitory areas of the underlying retinal ganglion cells. The antagonistic impulses that travel to the brain signal the existence of these illusory shadows.

Retinal ganglions appear to respond best to changes in light intensity created by borders. As a result, scientists speculate that the eye sees objects only by discerning edges. This rarely affects normal vision since people seldom encounter boundless objects of uniform color except in cleverly designed geometric illusions. Scientists discovered this aspect of the visual system by testing human perceptions of diagrams that trick the eye into seeing illusory edges. In one test, two quadrangles appear to be different shades of gray. In reality, the design is uniformly intense except for an abrupt change in brightness at the center. Covering this apparent edge with a pencil reveals that the two quadrangles are the same shade of gray. Without the illusory edge, the two distinct forms completely disappear. Investigators surmise that retinal ganglion cells detect only the transition in brightness at the center of the design and report this illusory edge to the brain.

The visual cortex falsely interprets the information and "sees" the change in intensity spreading to each side from the apparent transitional zone.

An array of commonly experienced illusions called afterimages also originate in the retina. It is not unusual to glance at a bright light and discover that a dark spot briefly remains as if it were painted on the retina. Some scientists attribute this afterimage to the massive chemical reaction that takes place in the retina as photoreceptors respond to the surge of light. Normally, the recovery process takes only a fraction of a second, but when the light is very intense, or occurs in a short burst, the receptors need more time to recover and readjust for normal light sensitivity. The illusion is more precisely called a negative afterimage because the light image appears dark.

Orientation and movement aftereffects seem to work in a similar fashion, but on different levels in the visual system. By studying people's responses to patterns of lines tilted at different angles, scientists have concluded that certain nerve cells, called line detectors, respond best to specific angles and directional movements. Suppose a person looks for three minutes at a line tilted fifteen degrees to the right. Nerve cells sensitive to that angle rapidly grow fatigued. When

Stare at the green pattern, below left, for ten seconds, then at the red-striped box for ten seconds. After three minutes of alternate viewing, gaze at the "colorless" design, opposite. The black-and-white pat-

tern now appears tinted. Turning the page 90° will cause the colors to reverse. Experiments have shown that this afterimage, called the McCollough effect, can linger for weeks. Science still does not fully

understand the illusion. Researchers believe the McCollough effect is more complex than most afterimages because the red vertical lines switch to horizontal and the green lines become vertical.

the observer shifts his attention to a vertical line, the balance of neural chemical activity switches to line detectors that respond best to lines slanted in the opposite direction. Consequently, the perpendicular lines appear to be tilted fifteen degrees to the left.

Neural adaptation might also underlie movement aftereffects. The most dramatic example is the waterfall phenomenon — so named for the illusion that results after a prolonged viewing of cascading water. When a person gazes at a waterfall for three minutes or longer, neurons that detect downward motion become fatigued by unrelenting stimulation. The resulting imbalance of neural activity causes neurons sensitive to the opposite direction of motion to take over. When the viewer looks away from the waterfall, stationary objects appear to sail upward although they do not seem to change location. This perceptual paradox demonstrates that parts of the visual system can succumb to illusions while other pathways of vision remain unaffected.

Sometimes the visual system detects shapes although images are not reflected onto the retina. These illusions, known as subjective contours, can be shown in the form of three black circles with pie-shaped wedges cut out. The "pies" are

arranged on a white background, so that they form the three corners of a triangle. When an observer views the pattern, he sees an illusory opaque triangle obscuring three circles that recede into the background. Concentrating on the circles causes the subjective contours to vanish.

While a number of theories have been proposed to explain subjective contours, none has gained general acceptance. One theory asserts that the background shapes stimulate neurons which detect corners. The brain interprets these signals as a continuous line. Yet, this does not explain why the illusion remains when dots replace the geometric shapes.

According to other theories, subjective contours deceive thought processes. Some scientists speculate that the mind organizes the incomplete background figures into stable, regular forms by imagining the presence of a superimposed image. In the subjective triangle illusion, the three sets of circular lines acquire completeness and regularity when perceived as disks obscured by an imaginary opaque form. Since the triangle must logically have a border, the mind somehow supplies the necessary lines.

Subjective contours add a new twist to the conventional geometric illusions continuing to

The illusory contour, right, forms an opaque triangle that exists only in the mind.

Diagonal lines in Zöllner's illusion seem to converge, but they are really parallel.

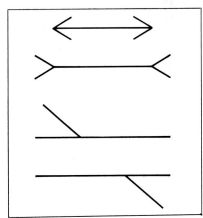

The Müller-Lyer arrows, top, and the Poggendorf illusion trick eye and brain.

baffle scientists. Despite more than a century of investigation, researchers still are not certain whether interactions among the visual system's neurons or mistaken mental judgments cause most illusory effects. One illusion that may arise from neural interaction is the pattern German scientist Johann Zöllner first noticed on a piece of fabric in a dress shop. When broadly viewed, pairs of parallel diagonal lines seem altered by slatted lines that give the figure the appearance of a diverging herringbone pattern. But the illusion will sometimes vanish if the observer continues to stare at it. One physiological theory proposes that the perceived slant of a line is determined by the "peak activity" of line-detecting cells, each of which fires in response to a specific range of positions. Line detectors responding to a narrow range of positions are called simple cells. Neurons sensitive to a wider range are known as complex cells. If complex cells err in response to the design while simple cells fire correctly, the contradictory signals would shift the lines' positions and cause the illusion.

Physiological theories cannot explain all geometric illusions, however. In the Poggendorf illusion, a pair of straight lines appear misaligned when they obliquely intersect two parallel lines.

Converging railroad tracks give depth cues similar to those that distort scaling in the Ponzo illusion. The two white lines are equal in length, but the top line appears to span more of the track than the one below it and so appears longer. Some scientists believe this "inappropriate size constancy" generates the illusion of unequal length.

The illusion persists even when illusory contours replace the parallel lines. Since nonexistent lines do not present images that trigger line-detecting neurons, investigators believe something other than neural activity is involved.

Perhaps the most familiar geometric illusion is the Müller-Lyer pattern. Two arrows with shafts of equal length appear to be different lengths when one has a projected point and the other, an inverted point. Research indicates that viewers compare the whole figures, including the points, to make their judgments of comparative size. According to one controversial theory, the line patterns are perceived as three-dimensional contours. The inverted points are depth cues, suggesting a perspective drawing of a far corner of a room. The other figure takes the shape of a near corner. The form resembling the far corner is more distant and so must be bigger. This theory of "inappropriate size constancy" also applies to the Ponzo illusion in which two lines of the same length appear unequal when presented against a background of converging lines. Many experts believe the background lines are seen as a perspective drawing of railroad tracks receding into the distance. The horizontal bar closest to the point of convergence would be more distant and, since it spans more of the track than the closer line, it appears to be larger.

A Dual Reality

Today, drawings that indicate depth are taken for granted, but in Renaissance Italy they were a novelty that amazed onlookers. For it was not until artists mastered scientific perspective in the early fifteenth century that man convincingly represented depth on a flat surface.

In 1428, the streets of Florence echoed with talk of a remarkable painting in the Church of Santa Maria Novella. The fresco, painted by a twenty-six-year-old artist nicknamed Masaccio, "Ugly Thomas," showed a somber couple praying outside a corridor. Beneath them, a skeleton lay on a sarcophagus. The figures of the Trinity appeared in the passageway behind the kneeling figures of Mary and St. John. The corridor, a great vaulted room flanked with massive columns, evoked a sense of spaciousness and depth

that awed the onlookers. The artist's brush seemed to have dissolved the wall, revealing a hidden chamber behind it.

The inscription above the skeleton, "I was once that which you are, and what I am you also will be," proved to be more than an epitaph. The painting was one of the first to follow the mathematical laws of linear perspective, determined by a famed contemporary of Masaccio's, Florentine architect Filippo Brunelleschi. The laws of perspective revolutionized art. What was it about Masaccio's painting that made people reach out to touch the plaster wall as if to test its solidity? The painting's success lay in the technical ability of the artist and in the skill — or weakness — of the visual system in perceiving what was not really there. Artists use this natural ability of eye and brain to create alternative realities. "Pictures are perhaps the first step away from immediate reality," notes Richard Gregory, a specialist in perception of art and illusion, "and without this reality cannot be deeply understood."

The study of how man perceives an illusory third dimension begins with the way he perceives a real third dimension. Within a range of about 200 feet, each eye sees a slightly different view of the surroundings. When combined, the eyes create a scene that has depth and distance. A man with only one eye, or someone viewing an object more than 200 feet away, does not use binocular vision. Instead, he unconsciously relies on cues that infer depth and distance. His brain, acting as a detective, assembles clues from the scene and combines them with what the man already knows to provide a nearly infallible report of the scene. That it goes unnoticed is the sign of this mechanism's success.

When a person studies a painting to determine how it was crafted, his eyes deliver the information he needs to study the way the canvas rests inside the frame, the thickness of the paint and the way the painting is mounted on the wall. But when he focuses on the scene depicted in the painting, the viewer employs the same cues for depth and distance and is susceptible to the same errors as a one-eyed man.

Art contains a built-in paradox, a dual reality. While a one-eyed man needs monocular depth

137

The artist's skilled use of depth cues creates a wondrous deception. In Giovanni Paolo Panini's mid-eighteenth century work, Interior of Saint Peter's, Rome, *grand corridors beckon the viewer.*

Distorted depth cues can cause visual chaos. "Whosoever maketh a design without the knowledge of perspective will be liable to such absurdities," wrote artist William Hogarth of his satirical 1754 engraving.

and distance cues to achieve a more complete picture of reality, the same cues in art achieve only illusion. A painting exists as two separate objects. It is the flat surface upon which paint is applied. And it is also the subject that appears in three dimensions on a two-dimensional canvas — the craggy face of an old man, a light-filled room, a violent battle scene. When an observer concentrates on a scene, he succumbs to the illusion of depth and forgets that he is looking at a two-dimensional surface.

Guide and Gateway

Linear perspective, the great preoccupation of Renaissance artists, is a key tool for creating the illusion of a third dimension. Leonardo da Vinci called it "the bridle and rudder of painting." Converging rows in a field of corn provide a clear example of linear perspective. The rows of corn appear to merge as they move upward on the visual plane. This gives an impression of distance. If one can see far enough, the vanishing point is visible — the place where corn, sky and horizon meet in a haze that melds them together.

Leonardo's *Mona Lisa* contains an inventive experiment in perspective. The famous portrait has two vanishing points, one for the woman's figure, the other for the scenery. "Those who are in love with practice without knowledge are like the sailor who gets into a ship without rudder or compass and who never can be certain whether he is going," Leonardo wrote in his notebooks. "Practice must always be founded on sound theory, and to this Perspective is the guide and the gateway . . . in the matter of drawing."

Landscape paintings usually contain many monocular cues to depth. In a picture of sheep grazing on a hillside, the sheep in the background are spots of white, while those in the foreground are larger and finely detailed. The viewer's prior knowledge of sheep — however minimal — aids him in seeing them as the artist intended. He knows the true relative sizes of the objects in the painting — sheep, trees and hills — so all appear normal in size.

Unless he has chosen an unusual perspective, the artist paints a scene from a slightly raised angle because his eyes are higher than the visual

plane. To enhance the feeling of depth, the artist places distant objects higher on the painting. One sheep grazing twenty feet behind another appears slightly above the first. A distant clump of trees is even higher on the canvas and the final mountain range is highest of all. If one mountain partially obscures the view of another, it is immediately interpreted as the nearer of the two, increasing the artist's illusion. The artist also adds a hazy blue tint to the distant mountain range, imitating an atmospheric effect of nature. To heighten the effect of distance, the artist evens out textures of distant trees, grass and the curls of the sheep's woolly coats. The rough surfaces of the mountains are smoother in texture the more distant they are from the viewer. The artist also models figures in light and shadow to create three-dimensional forms.

While some artists use distance cues to create an illusory third dimension, others manipulate

the cues to create a fanciful, distorted world. When eighteenth-century English artist William Hogarth designed a print for the title page of a textbook on perspective, he warped as many rules of perspective as he could. Size cues were reversed and inconsistent. Tiny sheep appeared in the foreground, while larger ones were placed farther away. Round barrels were incorrectly drawn. A distant hillside collided abruptly with a nearby house, and other images jostled one another in humorous confusion.

Nowhere is the visual paradox of art more apparent than in the work of M. C. Escher. The Dutch artist's sophisticated prints and drawings are at once rational and impossible, solid and insubstantial. They create worlds that can exist in only two dimensions. His renderings are so precise that the eye first willingly accepts them as rational. But once the mind analyzes the scene, the viewer realizes these never-ending staircases,

these pictures in which up and down have no meaning, are something not of this world but, perhaps, of another.

Deceptions to Delight

Just as we delight in Escher's sorcery, we enjoy being deceived by the illusions conjured by magicians. The use of mirrors and crude slide projectors called magic lanterns were as much a part of early magic acts as distracting words and sleight of hand.

In 1797, Belgian performer Etienne Gaspard Robert founded a theater of magic in the abandoned chapel of an old Capuchin monastery. His theater was so dimly lit that stealthy assistants could easily elude the audience. Robert tossed chemicals onto burning coals, causing columns of thick white smoke to rise in the dark theater. Meanwhile, assistants behind the stage lit a magic lantern which projected images of phantoms

M.C. Escher

Impossible Worlds

In his self-portraits, M. C. Escher's eyes radiate intense energy. Their magnetic gaze challenges the world from the paper. Yet, it is easy to feel that Escher directed his challenge mainly to himself, that he struggled more with the world behind his eyes than the world that lay before them.

The Dutch artist insisted his most brilliant images remained in his head, trapped there by his inability to render them on paper as vividly as they appeared in his imagination. "In comparison with these thoughts, every single print is a failure. . . ." he mourned. But with the images he put on paper — "very difficult and obstinate creatures" — Escher created an awesome and fascinating world, a place with the look of a fortress and the substance of a cloud.

As a student, Escher showed promise only as an artist. Yet the artistic community did not receive him warmly. Escher was "too tight, too literary-philosophical . . . too lacking in feeling or caprice, too little of an artist," concluded his professors at The School for Architecture and Decorative Arts in Haarlem.

Escher's early work was realistic. His superb technical skill gave his prints of landscapes and architecture an icy

clarity. Yet even in these early works, he experimented with unusual objects and angles of view. In the late 1930s, Escher's work underwent a sudden and complete change. He preserved his unique style but set it on a new course, into the realm of illusion. "I discovered that technical mastery was no longer my sole aim," Escher recounted. "Ideas came into my mind quite unrelated to graphic art," he wrote, "notions which so fascinated me that I longed to communicate them to other people. This could not be achieved through words, for these thoughts were not literary ones, but mental

images of a kind that can only be made comprehensible to others by presenting them as visual images."

Awkwardly at first, but then with increasing skill, Escher refined his ideas. Still, his art did not win wide approval. Critics considered his work more cerebral than artistic. Its popularity in scientific circles only served to underline such criticism.

Clearly, Escher understood the mechanisms of visual perception, for he could so skillfully lead eye and brain astray. One of his most famous prints showed a pair of hands, apparently rising out of their two-dimensional paper world, each a product of the other's pencil. His pen built impossible structures on the page that appeared plausible until the eye traced their corridors.

Escher's work showed that the cues the brain uses to make sense of the real world can form a nonsensical world just as easily. His fanciful world adheres to a different set of rules, like the Wonderland Alice stepped into through the looking glass. Escher knew well the boundary between the two worlds, for he felt a stranger to one of them. "I have no idea how to cope with reality," the artist once said, "my work does not touch it."

and spirits against pillars of smoke. Since no depth cues were available, the audience shrieked, convinced that the terrifying apparitions were only a few feet away.

Magicians also bounced images off mirrors. In the famous sphinx illusion, the conjurer placed a box on top of a table. When he opened the front panel, a sphinx's head appeared. The head belonged to the magician's assistant who was kneeling beneath the table. Finely polished mirrors concealed the table's flanks and, reflecting the image of the floor and side walls, hid the assistant from view. The audience would "look through" the mirror and see what they thought was the back wall. Thus hidden, it was a simple matter for the assistant to thrust his head out of a trap door in the table top and through the false bottom of the sphinx's box. The act fooled spectators for many engagements until a careful observer noted fingerprints on the mirror and discovered the illusion.

With the invention of the motion picture projector, magicians' illusions became even more startling. "Living pictures" became integral parts of magic acts. Today, cinema and television audiences take the illusion of apparent movement for granted. Yet scientists still do not fully understand how the visual system detects real motion. Apparent motion in movies and television is even more puzzling. Researchers study the illusion by using a machine that flashes two or more lights sequentially. If the lights are flashed at correct intervals, the viewer will see a single light moving across the screen. Unlike true motion, which occurs when an image sweeps across the retina, apparent movement takes place when an image jumps a gap between photoreceptors in the eye. This phenomenon provides the basis for viewing movies in which still photographs are projected on a screen at a rate of up to seventy-two flashes a second. The screen is dark half the time, but the images appear so rapidly that photoreceptors in the eye continue to fire between frames. In television, images flash sixty times a second, in consecutive bands that build up the picture in horizontal bars.

Motion pictures became more sophisticated with the advent of three-dimensional movies. To create the effect of depth, cinematographers record scenes in red and green or with polarized film at angles that mimic the different views both eyes ordinarily see. The audience receives spectacles with one red and one green lens or dual polarized lenses. The spectacles separate the overlapping images and create the illusion of figures and objects floating in front of and beyond the surface of the screen.

Strategy for Survival

While man tricks the eye for entertainment, insects and animals create illusions to survive. Some species of *Argiope* spiders in Brazil weave webs with intricate zigzag patterns resembling geometric illusions. The design so deceives the eye that it is difficult to find the spider in the center of the web. The wing cases of many tropical beetles diffract light in such a way that the beetle's size and shape appear to change depending on the angle at which it is viewed. Predatory birds cannot easily estimate the beetle's distance.

Some fish that live in the depths of the sea emit light through organs called photophores. The *Argyropelecus affinis*, or hatchet fish, casts light toward its shadow through a system of lenses and reflective tissue. This makes it difficult for

When disturbed, the inchworm caterpillar, on the right, clutches a branch with its hind legs and stands erect. Perfectly still, the worm closely resembles the small twigs found on the trees upon which it feeds. Many scientists believe the inchworm developed this ruse to trick the eye of its predators.

With its hairy back, the black-and-yellow striped robber fly, bottom, mimics the bumblebee, top. Thus disguised, the fly easily approaches the unwary bumblebee upon which it preys.

predators to track the fish from below. According to British researchers, the fish possess color filters that modify the light produced by the photophores until it matches the wavelength of sunlight penetrating to the depth at which the fish live. When the hatchet fish moves closer to the surface, its photophores emit light of a higher intensity to match the level of light surrounding it. *Chauliodus sloani*, a fish that seldom leaves deep water, has smaller photophores that produce a weak light effective for camouflage.

Many scientists believe insects and animals have learned to mimic not only the appearance but also the behavior of harmful species to ward away predators. Probably evolutionary in origin, this ruse enables moths and other harmless butterflies to imitate the poisonous *Danaus chrysippus*, including the white spots on its red wings. The *Dirphya* beetle not only resembles a Braconid wasp, it also behaves like one. When provoked, the beetle threatens to sting its attacker even though it has no stinger. When British scientist Geoffrey Carpenter picked up the *Dirphya,* it "curved the top of its abdomen . . . and actually protruded a flexible white viscus which it moved about just like a sting." The beetle's false sting "was so very striking that, although reason told me it was a beetle, instinct was so strong that misgivings almost prevented [my] handling it," Carpenter said.

Female *Photuris* fireflies mimic light signals that serve as a mating call for *Photinus*, another species of firefly. The length of the female's light flashes, the number of pulses and the flash rate normally enable the male *Photinus* to distinguish the female of the species. The female *Photuris* uses bogus signals to attract *Photinus* males which respond to the call and attempt to mate with the firefly. Instead, the *Photuris* female devours the unwary male. When the *Photuris* female mates, she reverts to the system of light signals that attracts males of her own species.

Some scientists believe the developmental stages of certain butterflies mimic the faces of monkeys. The African *Spalgis lemolea* resembles the *Cercopithecus* monkey while pupae of oriental butterflies look like monkeys found solely in the Far East. Such mimicry, scientists speculate, serves to ward off birds which hunt by a method called rapid peering. The birds inspect objects from several angles in rapid succession at very close range. Their binocular field is thought to be so narrow that the apparent distance of a familiar object is determined solely by the size of its image upon the retina. When one of these birds finds the pupae of the blue butterfly, it suddenly confronts what looks like the face of a monkey and flies away.

The Bizarre in Space

In leaving the planet and venturing into space, astronauts encounter bizarre illusions, unlike anything man has confronted on Earth. Such illusions might spell disaster during critical flight operations demanding unerring sight. Absence of normal distance cues in the vacuum of space makes size and depth perception unreliable. On Earth, the eyes focus on distant scenery through progressive adjustments to a series of distant objects. During space flight, however, the astronaut does not have a succession of objects to determine distance. Within the space capsule, his eyes focus on a console of instruments close at hand. Consequently, he might not be able to determine whether his eyes are focusing at twenty feet or at infinity when he looks outside the craft. He might miss objects that are only one hundred feet away. In space, depth perception and visual acuity depend on stereoscopic vision — the ability to see objects in three dimensions — and on the available depth cues such as the absolute size of the moon, Earth and manmade satellites. However, once space voyagers travel beyond the Earth and moon, these depth cues will no longer be available.

On earth, light scatters from molecules of air, dust and moisture. In space, the distribution of light meets no resistance. This enables the astronaut to see contrasts of light and darkness more clearly than is possible on earth. While this means the human eye may have greater acuity, increased contrast perception can distort shapes. The eye might perceive sharp shadows cast against an orbiting space station as the edges of actual objects. Such illusions can influence judgments of size and distance during rendezvous.

Lack of depth cues in the vacuum of space makes Dione appear much larger than the planet she orbits in the montage of the Saturnian system, above. Nor can the unaided eye tell whether the other celestial bodies are Saturn's moons or distant planets. Starlight passing through the gas of the Orion Nebula, opposite, can cause errors in calculating stellar distances just as fog hinders motorists' perception of depth.

With prolonged exposure to negative gravity, astronauts have experienced "redout," a reddish haze caused by congested blood vessels in the eyes. Gravitational forces during launch have also caused "grayout," the temporary dimming of central vision. Astronauts on the orbital space station, Skylab, tested their susceptibility to visual illusions experienced in preflight tests. In one experiment, astronauts donned goggles that presented a star of light to the left eye. During clockwise rotation, many astronauts reported that the star oscillated. Scientists found that rotation caused quick eye motion in the direction of revolution followed by slow eye movement in the opposite direction. During the quick saccadic movements, the star passed across the retina. The astronauts, unaware of the eye movement, attributed the illusory movement to the star.

In another test with the same apparatus, astronauts recalled that during acceleration they felt as if they were slowly tilting backwards. At the same time, the star appeared to rise until it hovered directly above them. The illusion is so persuasive that inexperienced subjects often believe the illusion is real. Technicians guard against the autokinetic effect — an illusion making a small light source wander aimlessly against a uniformly dark background — by providing strong illumination within the space craft.

During the Apollo space flights of the late 1960s, astronauts reported seeing luminous streaks and bright spots of light before their eyes. Ground crew scientists attributed the illusions to the ship's magnetic field, designed to protect the occupants from the sun's radiation during solar flares and sunspot storms. Too much oxygen in early space capsules produced tunnel vision, distorting the astronauts' peripheral vision into a swirl of obscure shapes. Because space travel can trick man's eye in so many ways, scientists train astronauts and jet pilots not to trust their eyes, but to rely on celestial guidance systems that report the craft's position and orientation. Still, man has not yet developed a guidance system that can replace the eye. When man landed on the moon, it was the trained eye of astronaut Neil Armstrong that guided the Eagle to a safe landing on the Sea of Tranquility.

Appendix

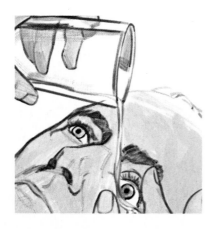

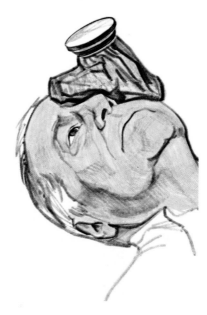

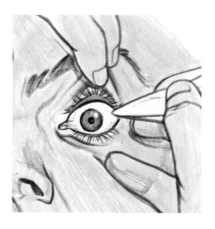

Eye problems range from simple eyestrain to major trauma. Cool compresses or rinsing with cool, clean water can often relieve tired, red eyes. But persistent redness can indicate a more serious problem. Generally, you should consult an ophthalmologist if you have any of the following problems:

1. Sight is impaired.
2. The eye becomes painful.
3. Foreign matter enters or punctures the eye.
4. The eyelids stick together.
5. Bright light is annoying.
6. The eyes remain bloodshot for several days without apparent reason.

When an emergency strikes, prompt first aid followed by professional medical treatment can often prevent permanent eye damage.

A black eye is a potentially serious injury that should receive medical attention. Cold compresses often relieve the initial swelling if the eye and its bony cavity are normal. Medical help should be summoned immediately if a fracture is suspected, or if the victim reports double vision. Thorough eye examinations can sometimes detect tiny rips in the retina that can develop into serious eye problems.

The eye traps foreign matter, but usually washes the particle out with tears. If the substance remains, or appears to have penetrated the eyeball, medical help should be summoned. The doctor will often bathe the cornea with steam or attempt to dislodge the substance with sterile gauze. The physician might then treat the condition as he would a scratch or cut on the cornea by applying an antibiotic solution and covering the eye with a patch until the eye heals.

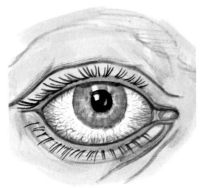

Adapted from CIBA-GEIGY.

Conjunctivitis is a common affliction caused by a variety of fungi, bacteria, allergens or chemicals. Commonly called pink eye, the condition results in inflammation and redness of the conjunctiva, the mucous membrane that covers the surface of the sclera. Viral conjunctivitis can cause irritated eyes accompanying colds. Often, the conjunctivitis sufferer awakens to discover his eyelids are stuck together by thick discharge. Antibiotic eye drops usually clear the condition.

Ninety percent of all eye injuries can be prevented by wearing protective eye gear and following a few simple safety rules, according to the National Society to Prevent Blindness. It has been estimated that certified face protectors for amateur hockey players avert 70,000 eye injuries a year and save more than $10 million in medical expenses annually. Impact resistant eye guards can protect against the force of a pebble thrown by a lawn mower or a smashed tennis or racquet ball. Safety goggles should also be worn when using power tools.

Overexposure to sun lamps or solar rays can result in serious eye injuries. Symptoms begin to develop within twelve hours. Often, a person who spent an afternoon sunbathing awakens during the night with the harrowing sensation of sand grains in his eyes. Mild analgesics can relieve the pain. Swelling will subside after twenty-four hours with prompt medical attention. But sight-threatening corneal scars sometimes remain. Ultraviolet burns can easily be avoided by wearing sun goggles when using sun lamps and donning dark sunglasses when lying in the sun.

Adapted from CIBA–GEIGY.

A sty is a painful infection of the glands between the lashes of the eyelid. Warm compresses will usually drain the small yellowish swelling. A chalazion develops in the same way, but often produces a large cystic swelling of the eyelid. The treatment is the same as for sties, although the doctor must often incise the chalazion in order to drain the lesion.

Chemicals can burn the eye and cause permanent damage unless the eye is immediately washed with clean water or a sterile saline solution. If the chemical was alkaline, the affected eye should be flushed for at least twenty minutes; if acidic, cleansing should last at least ten minutes or until medical help arrives.

Any cuts to the eye and eyelid can be serious. A clean, loose dressing should be applied to both eyes and medical help summoned immediately.

Any foreign matter protruding from the eye should not be removed. A paper cup should be placed over the wound and the victim kept lying quietly on his back.

In serious injuries, an eyeball may pop from its socket. It should not be pushed back. Cover both eyes with a moist, clean dressing and place a protective cone over the protruding eyeball. Do not apply pressure. The victim should lie on his back while waiting for professional medical care.

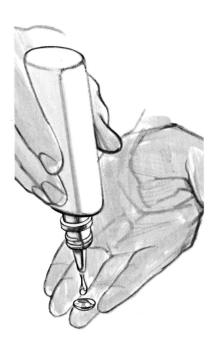

Daily care of contact lenses can help prevent corneal abrasions and infections. Lenses should be soaked periodically in cleaning solution. Before applying the lenses, use a wetting solution to help natural tears circulate around the lenses. Many ophthalmologists recommend that lens wearers store lenses overnight in a soaking solution to keep the lenses moist and clean.

Eye make-up can contain ingredients that irritate the eyes. Some cosmetics firms offer hypoallergenic products which they claim can reduce instances of irritation. Always be careful when using mascara or other make-up for coloring eyelashes or eyelids since negligent use of applicators can cause dangerous corneal abrasions and other eye injuries.

Smog and chemicals can irritate the eyes. Decongestant eye drops appear in a variety of over-the-counter products. To apply eye drops, pull the lower eyelid down to form a pocket. While looking up, place a drop in the pocket. Glaucoma sufferers should consult a physician before buying eye medication since many drugs can dilate the pupil and precipitate attacks.

Glossary

accommodation the automatic adjustment of the curvature of the eye's lens by contraction or relaxation of ciliary muscles to bring images from various distances into sharp focus on the retina.

adaptation the adjustment of the eye to changes in light intensity or color tone; adjustment of the eye's photoreceptor cells to a visual stimulus.

aftereffect a delayed or protracted physical or cognitive response to a stimulus, usually involving motion or orientation.

afterimage a visual picture that briefly remains after adaptation to an image.

amacrine cells modified nerve cells found at the eye's retina.

amblyopia dimness of the field of vision that occurs without apparent ocular impairment.

amplitude height of a wave's peak.

analgesic medication that reduces pain.

angiogram an X-ray picture of blood vessels that can reveal disorders such as blood clots and tumors.

aniseikonia a disease that causes the size and shape of a retinal image of an object to differ in each eye.

anomalous trichromacy a defect in color vision involving the perception of primary colors.

anterior chamber the space between the cornea and the lens that is filled with aqueous humor.

apparent distance theory an explanation for the illusion of the moon's apparent change in size when over the horizon and when high in the sky. It holds that the impression of distance created by terrain affects the perceived size of the moon.

apparent motion perceived movement of an image without actual movement.

aqueous humor the transparent fluid that fills the anterior and posterior chambers of the eye and that diffuses into the blood.

artifice a device used to create an illusory effect in art.

astigmatism an unequal curvature of the eyeball that results in distorted focal points of light on the retina.

autokinetic effect the illusion of movement generated by a single luminous object when viewed against a uniformly dark background.

axon the nerve cell's process which carries impulses away from the neural body.

bifocal lens a corrective lens for both near and distant vision.

binocular pertaining to both eyes.

bipolar cells retinal neurons associated with the cones or rods.

blind spot a small portion of the retina that is insensitive to light; located at the point at which the optic nerve passes through the back of the eyeball.

brightness the effect or sensation by which differences in luminance are distinguished.

camera obscura a box, lens and screen used for viewing images.

canal of Schlemm a channel at the passageway between the sclera and cornea that serves chiefly to drain aqueous humor from the eyeball.

canthus the angle formed by the meeting of the upper and lower eyelids at each corner of the eyes.

capillaries the smallest blood vessels.

cataract opacity of the eye's lens capsule or crystalline lens.

cell body the main portion of a neuron containing the nucleus.

cerebellum the twin-lobed, oval structure behind the brainstem responsible for coordinating movements.

cerebral cortex the thin, convoluted outer layer of the cerebral hemispheres, rich in nerve cells, and the birthplace of man's higher mental functions.

chalazion a swelling of the eyelid caused by the retention of meibomian gland secretions.

choroid the thin, vascular coat located between the eye's retina and the sclera.

ciliary body a ring of tissue that connects the choroid to the iris and aids in accommodation.

ciliary zonule the fibrous system that holds the eye's lens in place.

colliculus the elevated portion of the optic nerve at the point of entrance to the retina.

colorblindness defective vision involving faulty or deficient perception of color.

color circle an ordering of colors in terms of their comparability.

complex cortical cells line detectors; neurons believed to respond to a specific orientation of stripes of differing widths or to movement of lines in certain directions.

cones the light-sensitive retinal cells that respond best to bright light and provide detection of spectral colors.

conjunctiva the mucous membrane that lines the eyelids and covers the surface of the sclera.

conjunctivitis pink eye; inflammation of the conjunctiva.

contact lens a thin lens of glass or plastic fitted over the cornea to correct visual defects.

contraction the shortening of the eye muscles.

contrast sensitivity the limits to which an observer can distinguish differences in luminance.

convergence the concurrent inward movement of the two eyes toward a near object.

cornea the transparent anterior structure of the eye's outer coat that covers the iris and pupil.

corneal endothelium the layer of cells found in the eye's anterior chamber that covers the rear portions of the cornea.

cortex the gray matter on the outer surface of the brain.

couching the dislocation of the lens by surgical means.

cryoprobe a device that reduces the surface temperature of tissues to extremely low levels.

dendrite one of the threadlike extensions of a neuron that carries impulses toward the neural body.

deuteranopia defective color vision that distorts the perception of red and green.

dichromatic able to distinguish only two colors.

dilation widened or enlarged.

diopter a measurement of a lens's refractive power to focus on an object one meter away.

diplopia double vision; the visual perception of two images from a single object.

epithelium the tissue that covers internal and external body surfaces.

evil eye a glowering look believed by the superstitious to bring calamity upon the recipient.

fata morgana a mirage named for the nymph who in Arthurian legend created castles in the air.

Fechner's paradox the phenomenon in which the average level of illumination affects the size of the pupil.

fixation the act of training one's eyes on an object.

fluorescein a compound that fluoresces under blue or ultraviolet light; used to define areas of the eye to detect ocular defects.

focal length the distance from the point of focus to the surface of the lens.

focus the point at which light rays diverge or converge to form a clear and distinct image.

fovea centralis a small rodless pit in the macula lutea where visual perception is most acute.

frontal lobe one of four regions of the cerebral cortex. The frontal lobe lies directly behind the forehead. The hindmost part of this lobe contains the motor cortex, but the function of the remaining portion is unclear. Some researchers theorize that foresight, planning and personality are located here.

glaucoma an eye disease characterized by abnormal fluid pressure within the eyeball resulting in distortion of the optic disk.

gonioscope an optical instrument used to examine the angle of the eye's anterior chamber.

grayout the temporary diminution of central vision reported by astronauts during liftoff.

hallucination delusion; a false perception without an appropriate stimulus.

horizontal cells neurons, inhibited by the retinal rods and cones, which transmit inhibitory signals chiefly to bipolar cells.

hue the subjectively perceived quality of a color.

hyperopia farsightedness; a visual defect or refractive error that focuses light rays behind the retina due to an abnormally short eyeball.

illusion a systematically incorrect or misinterpreted sensory impression.

image the optically formed representation of an object.

inferior oblique muscle the musculature that turns the eye in upward and outward directions.

inferior rectus muscle the musculature that moves the eye downward.

iridectomy surgical removal of a portion of the iris.

iris the pigmented diaphragm which is perforated by the pupil and is located behind the cornea.

Klüver-Bucy syndrome psychic blindness; a phenomenon sometimes observed in persons suffering damage to the temporal lobes in which the ability to recognize familiar objects by sight is lost, often accompanied by a loss of fear of approaching objects that were formerly regarded as harmful.

lacrimal duct the passageway located behind the lower eyelids that drains used tears out through the nasal cavity.

lacrimal gland a spongelike organ that secretes tears.

lateral geniculate nucleus a region of the brain located in the thalamus which receives signals from the optic nerve.

lateral inhibition a process by which retinal ganglion cells are suppressed from firing.

lateral rectus muscle the system of contractile fibers that control the outward directional movements of the eye.

lens a transparent structure that is convex on front and back, located between the vitreous humor and the iris, and that focuses light rays onto the retina.

light electromagnetic radiation that the normal eye perceives as a source of illumination.

limbic system a group of structures at the base of the forebrain involved in emotions and behavior.

linear perspective an artistic technique that represents depth through real or implied lines converging in the background to delineate relative size and distance.

luminous emitting light.

lysozyme an antibacterial enzyme found in tears.

macula lutea an area near the central part of the retina composed almost completely of cones at which visual perception is highly acute.

manometer a gage for measuring the relative pressure of liquids and gases.

medial rectus muscle musculature that pulls the eye inward.

meibomian gland an organ located between the hair follicles of the eyelid that produces semifluid secretions.

melanin the brown or black pigment found in the vascular layer of the eye between the retina and the sclera.

microaneurysm a minute bubblelike swelling in the wall of a blood vessel.

microelectrode a wire-thin conductor used either for sending electrical current into specific brain regions, or for recording the brain's electrical activity.

mirage an optical phenomenon generated by the refractive distortion of light by alternate layers of hot and cool air.

monocular pertaining to or having one eye.

moon illusion a phenomenon in which the moon appears larger at the horizon than when at its zenith.

myopia nearsightedness; a condition in which light rays focus at a point in front of the eye's retina due to an elongated eyeball.

nasal cavity the air-filled sinus in the skull extending from the nostrils to the pharynx.

neuron nerve cell; the basic conducting unit of the nervous system consisting of a cell body and threadlike projections that conduct electrical impulses. The axon, a long fiber, transmits impulses, while the shorter extensions called dendrites receive them.

neurotransmitter a substance that can excite or inhibit the firing of nerve cells.

nucleus the portion of a living cell containing genetic information which controls the cell's growth, metabolism and reproduction.

occipital lobe the lower portion of the cerebral cortex containing the visual cortex.

ocular decongestant eye wash; a medicated solution used to provide temporary relief from tired, red eyes.

opaque impenetrable by light.

ophthalmology the branch of medicine specializing in eye disorders.

ophthalmoscope a mirrored instrument with lenses used during eye examinations.

opponent-process theory a theoretical explanation of color vision holding that the mixture of four opposing primary hues creates all the colors of the visible spectrum.

opsin a colorless protein associated with the retinal rods and cones that joins with various forms of vitamin A to create light-sensitive pigments.

optic   of or pertaining to the eye.

optic chiasm   the region of the hypothalamus where optical nerve fibers from the optic nerve cross over.

optic nerve   the nerve carrying visual information from the eye to the visual cortex via the optic tract.

optic tract   the optic nerve fibers continuing from the optic chiasm to the visual cortex.

orbit   the bony cavity surrounding the eye.

parallax   the apparent change in an object's direction as a result of the change in an observer's position.

perceptual constancy   the perception of a stable world regardless of inconsistent and confusing stimuli; often used in regard to perception of size.

photochemical   associated with light's chemical properties.

photophore   a light-emitting organ.

photoreceptor   a specialized cell sensitive to light and containing photopigments.

pigment   any substance or matter that colors the body.

pons   a band of nerve fibers at the front of the brainstem bridging the left and right halves of the cerebellum.

posterior chamber   the vitreous-filled space of the eye between the back side of the iris and the lens and ciliary body.

presbyopia   diminution of focal ability caused by the loss of elasticity of the crystalline lens of the eye usually associated with age.

prestriate cortex   an area of the visual cortex located anterior to the striate cortex which is thought to be involved in the higher functions of visual perception.

primary colors   pure colors which generate all colors of the spectrum.

prism   a transparent object that produces or divides a spectrum of light.

process   a cellular appendage.

protanopia   a form of colorblindness in which the ability to perceive blue and yellow is retained.

Pulfrich effect   the phenomenon by which a pendulum that oscillates back and forth appears to swing elliptically

when viewed normally with one eye and through a dark filter over the other eye.

pupil   the central opening of the eye's iris.

quantum   a unit of energy.

receptive fields   areas that blanket the surface of the retina and which, when responding to a spot of light, affect the ganglion cell's behavior.

redout   the temporary distortion of vision by a reddish haze caused by congested vessels in the eye.

refraction   the bending of a light wave.

retina   the small patch of tissue at the back of the eyeball that contains the light-sensitive rods and cones.

retinal ganglion cells   a group of nerve cell bodies located at the retina forming the optic nerve.

retinitis pigmentosa   hereditary diseases characterized by progressive blindness.

retinoscope   an optical instrument for performing retinoscopy.

retinoscopy   a medical evaluation of the eye's refractive properties.

rhodopsin   visual purple; a light-sensitive pigment found in the rods which undergoes change in response to light.

rods   highly specialized cylindrical light-sensitive cells containing rhodopsin which are most sensitive to dim light.

saccades   quick, involuntary eye movements during changes in points of fixation.

saturated color   a pure color; vividness of hue.

sclera   the hard, white outer coat of the eye.

scotoma   a blind spot or area in which normal vision surrounds an area of depressed sight.

scotopic vision   night vision; perception of images under conditions of reduced illumination.

Snellen's chart   a chart of block letters of gradually increasing sizes, marked according to the distances at which they are normally legible; used in visual acuity tests.

spectrum   the distribution of energy in order of wavelengths.

stereoscope   an optical instrument that gives a three-dimensional effect to two

photographs taken at angles which mimic the disparate views that the two eyes normally see.

stereoscopic vision   the ability to perceive objects as three-dimensional.

strabismus   an ocular defect in which one eye cannot focus on an object due to eye muscle imbalance.

striate cortex   an area of the occipital portions of the brain distinguished by its striped appearance.

stroma   the eye's supporting tissue.

sty   a painful infection and inflammation of the glands between the lashes of the eyelid.

superior oblique muscle   the musculature that turns the eye downward and slightly outward.

superior rectus muscle   musculature that moves the eye upward.

temporal lobe   the area of the cerebral cortex located near the temples of the skull that contains centers for hearing and memory.

thalamus   a twin-lobed mass of nerve cells at the top of the brainstem containing relay centers for sensory and motor information to and from the brain.

trabecular meshwork   connective tissue that supports or anchors an organ.

transparency   the transmission of light in such a way that images can be seen as though no intervening material exists.

tunnel vision   a sight disorder characterized by loss of the peripheral field of vision.

vergence   the simultaneous movement of each eye in opposite directions.

visual acuity   sharpness of vision.

visual cortex   that part of the brain specializing in the interpretation of visual signals.

visual field   the area of sight.

visual system   the pathway of sight from eye to brain.

vitrectomy   surgical removal of clouded vitreous humor followed by its replacement with a clear, saline solution.

vitreous humor   the gel-like substance filling the vitreous chamber.

wavelength   the distance between corresponding points in two consecutive cycles.

zenith   the highest point in the celestial sphere.

Photographic Credits

So Great a Wonder
8, Navin Kumar Gallery, New York. 10, (top) © Sonia Halliday & Laura Luchington (bottom) Walters Art Gallery, Baltimore. 11, The Bettmann Archive. 13, (top) Scala Editorial Photocolor Archive (bottom) Photo Bibliothèque Nationale, Paris. 14, The Bettmann Archive. 15, (top) © Sonia Halliday & Laura Luchington (bottom) British Museum. 16, *Saint Lucy*, by Francesco del Cossa, National Gallery of Art, Washington, Samuel H. Kress Collection. 17, Collection of G. E. Mestler, Downstate Medical Center, Brooklyn. 18, National Library of Medicine. 19, (both) The Bettmann Archive. 20, **Thomas B. Allen.** 21, (top) The Bettman Archive (bottom) Bodleian Library MS WC 361 vol. 2. 22, (top) Mary Evans Picture Library (bottom) From I. Bernard Cohen, *Album of Science, from Leonardo to Lavoisier* (New York: Charles Scribner's Sons, 1980. 23, **Thomas B. Allen.** 24, (right) The Bettmann Archive (left) National Library of Medicine. 25, (both) Collection of G. E. Mestler, Downstate Medical Center, Brooklyn.

The Mechanics of Vision
26, Howard Sochurek/Woodfin Camp & Associates. 28, (top) Dan McCoy/Rainbow (bottom) **Ray Srugis.** 29, Manfred Kage/Peter Arnold, Inc. 30, Jane Burton/Bruce Coleman, Inc. 31, Manfred Kage/Peter Arnold, Inc. 32, from *Tissues and Organs: A Text Atlas of Scanning Electron Microscopy* by Richard Kessel & Randy H. Kador; W. H. Freeman & Company © 1979. 33, **Ken Goldammer.** 34, **Thomas B. Allen.** 35, T. Kuwabara, MD, National Eye Institute. 36, F. M. de Monasterio, S. J. Schein, E. P. McCrane, National Eye Institute. 37, (both) Pat Field. 38, College of Physicians of Philadelphia. 39, **Thomas B. Allen.** 40, 41, **George V. Kelvin.** 42, (both) William Vandivert and *Scientific American*. 43, **Adolf E. Brotman.** 44, National Library of Medicine. 45, **George V. Kelvin.** 46, Ken Touchton. 47, (top) Ken Touchton (bottom) **Esperance Shatarah.** 48, Telcom Research, Inc., Teaneck, NJ. 49, Dan Margolish, Cal Tech, Pasadena, CA.

Pictures in the Mind
50, FPG/Page. 52, Manfred Kage/Peter Arnold, Inc. 53, (top) **Susan Johnston** (bottom) **Esperance Shatarah**, adapted from Brodmann, K., *Vergleichende Lokalisations lehre der Grosshirnrinde*, Leipzig, Barth, 1909, pp. 144-45. 54, (top) Picture courtesy of Doctors Mortimer Mishkin, Charles Kennedy and Louis Sokoloff at the NIMH, Bethesda, MD (bottom, both) adapted from Ungerleider, L. G. and Mishkin, M., "Two Cortical Visual Systems," in *Advances in the Analysis of Visual Behaviour*, edited by D. J. Ingle; R. J. W. Mansfield; M. A. Goodale, M.I.T. Press, Cambridge, MA, 1981. 55, **Thomas B. Allen.** 56, (top) **Chip Coblyn** (bottom) **Pam Schillig.** 57, (left) **Louis**

W. Bory Assoc. (right) Zig Leszczynski/Animals, Animals. 58, (both) Dr. David Hubel and the Dept. of Neurobiology, Harvard Medical School, Boston. 59, **Peggy Gage/Karen Karlsson** (inset) Mickey Palmer/Focus on Sports. 60, Dr. H. V. B. Hirsch, State University of New York, Albany. 61, **Martha Anne Scheele.** 62, The Coca-Cola Company. 63, Jane Welliver, courtesy of National Exhibits by Blind Artists. 64, William Vandivert and *Scientific American*. 65, Sherwin Isenberg, MD, UCLA School of Medicine. 66, © 1979 Eddie Adams/Contact. 67, (top left) Dewitt Jones/Woodfin Camp & Associates (top right) Chuck O'Rear/Woodfin Camp & Associates (bottom) Dewitt Jones/Woodfin Camp & Associates. 68, (both) The Bettmann Archive. 69, **Susan Johnston.** 70, Michael de Camp. 71, courtesy of Kaiser Porcelain Limited of England. 72, redrawn by **Peggy Gage.** Based on drawings from R. L. Gregory and E. H. Gombrich, Editors, *Illusion in Nature and Art.* Copyright © 1973 by Colin Blakemore, Jan D. Deregowski, E. H. Gombrich, R. L. Gregory, H. P. Hinton, Roland Primrose (New York: Charles Scribner's Sons, 1974) reprinted with permission of Charles Scribner's Sons and Duckworth. 73, **Thomas B. Allen.** 75, © 1975 by Ronald K. Siegel.

Sensing Color
76, reprinted from *The Spirit of Color* by Karl Gerstner by permission of M.I.T. Press, Cambridge, MA. 78, **Another Color.** 79, (left) Harold Hoffman/Photo Researchers (top right) C. Roessler/Animals, Animals (bottom right) Thomas R. Taylor/Photo Researchers. 80, *Madonna and Child with the Baptist and St. John*, attributed to Cimabue, National Gallery of Art, Washington, Samuel Kress Collection. 81, National Gallery, London. 82, Fritz Goro/*Life* magazine © Time Inc. 83, (top) Georg Field, *Chrometrics*, London, 1817 (bottom) M.I.T. Library, Institute Archive and Special Collections. 84, **Another Color.** 85, **Thomas B. Allen.** 86, (left) **Another Color** (right) *Science* — fig. 1 from "Staining of Blue-Sensitive Cones of the Macauque Retina by a Fluorescent Dye," F. M. de Monasterio *et al.*, vol. 213, p. 1278, 11 September 1981 by the American Association for the Advancement of Science. 88, Kanaehara Shuppan Co. LTD, Tokyo. 89, Max Hirshfeld. 90, Courtesy Cavendish Laboratory, Cambridge, England. 91, Lee Foster/Bruce Coleman, Inc. 92, (top) Eric Leslie Simmons/The Image Bank (bottom) Alfred Pasieka/Bruce Coleman, Inc. 93, B. J. Spenceley/Bruce Coleman, Inc. 94, Michael de Camp. 95, **Chip Coblyn/Bethann Thornburgh.** 96, (both) Norton Simon Museum of Art, Pasadena. 98, (top) The Art Institute of Chicago (bottom) The Metropolitan Museum of Art. 99, The Phillips Collection, Washington.

Disorders of the Eye
100, Alexander Tsiaras/*Discover* © 1981 Time Inc. 102, © Ken Touchton. 103, © Muriel Laban Nussbaum, 1977. 104, **Another Color.** 106, The Bettmann Archive. 107, **Thomas B. Allen.** 109, (top left) Mitch Kezar/Black Star (top right) Dan McCoy/Rainbow (bottom right) © Ken Touchton. 110, Collection of G. E. Mestler, Downstate Medical Center, Brooklyn. ΟΦΘΑΛΜΟΔΟΥΛΕΙΑ, *Das Ist Augendienst*, Georg Bartisch, Dresden, 1583. 111, Bodleian Library, MS. Ashmole 1462, folio 10r. 112, 113, Alexander Tsiaras. 115, (both) © Muriel Laban Nussbaum, 1981. 116, © Ken Touchton. 117, (top) Dan McCoy/Rainbow (bottom) Alexander Tsiaras/*Discover* © 1981 Time Inc. 118, **Ray Srugis.** 119, Alexander Tsiaras. 120, Christopher Springmann/Black Star. 121, Lee Foster/Bruce Coleman, Inc. 122, 123, Alexander Tsiaras.

Illusion and Artifice
124, Collection Haags Gemeentemuseum - The Hague, © 1981 BEELDRECHT Amsterdam/VAGA, New York. 126, David Muench. 127, Keith Gunnar/Bruce Coleman, Inc. 128, © Dargis Associates, Inc. 130, (both) © Roxby Press Ltd. 131, From R. L. Gregory and E. H. Gombrich, Editors, *Illusion in Nature and Art.* Copyright © 1973 by Colin Blakemore, Jan D. Deregowski, E. H. Gombrich, R. L. Gregory, H. P. Hinton, Roland Primrose (New York: Charles Scribner's Sons, 1974) reprinted with permission of Charles Scribner's Sons and Duckworth. 132, (top) photo copyright © 1981 Arthur Sirdofsky (bottom) **Esperance Shatarah.** 133, (left) The Bettmann Archive (right) © Roxby Press Ltd. 134, (both) © Roxby Press Ltd. 135, (top left) © Roxby Press Ltd. (top right) reprinted with permission of Macmillan Publishing Co., Inc., from *An Introduction to Perception* by Irvin Rock (middle right) from R. L. Gregory and E. H. Gombrich, Editors, *Illusion in Nature and Art.* Copyright © 1973 by Colin Blakemore, Jan D. Deregowski, E. H. Gombrich, R. L. Gregory, H. P. Hinton, Roland Primrose (New York: Charles Scribner's Sons, 1974) permission, Charles Scribner's Sons and Duckworth (bottom right) **Esperance Shatarah.** 136, © David Plowden/Photo Researchers. 137, SCALA/Editorial Photocolor Archives. 138, *Interior of St. Peter's, Rome*, by Panini, National Gallery of Art, Washington, Ailsa Mellon Bruce Fund. 139, The Bettmann Archive. 140, Detail of Concave and Convex. Collection Haags Gemeentemuseum - The Hague, © 1981 BEELDRECHT Amsterdam/VAGA, New York. 141, **Thomas B. Allen.** 142, From Erik Barnouw, *The Magician and the Cinema* (New York: Oxford University Press, 1981) courtesy of Erik Barnouw. 143, John R. MacGregor/Peter Arnold, Inc. 144, (both) Carroll W. Perkins. 145, Zig Leszczynski/Animals, Animals. 146, 147, NASA.

Index

Page numbers in bold type indicate
illustrations and photographs

A
aberration, 30
accommodation, 32, 65, 69, 74, **104**, 105
adaptation, 37, 42, 43, 134
aether, 22, 23
afterimage, 15, 133, 134
Albers, Josef, **96**
Alhazen, 14-15, 17
al-Kindi, Ya'qub ibn-Ishaq, 14, 15
amacrine cells, **40**, 43, 51, 87
amblyopia, 61, 109
Ames, Adelbert, Jr., 130-31, **132**
amino acid, 54, 58
amygdala, 54
aniseikonia, 130
Annis, Robert, 72
anterior chamber, **33**, 102, 121
apparent distance theory, 129
applanation tonometer, 102, **103**
aqueous humor, 29, 30, 31, 114
Argus, **11**
Aristotle, 12, 14, **14**, 15, 17, 80, 82, 83
atomic theory, 11-12
Avicenna, 14, **14**
axon, **57**

B
Bacon, Roger, 15, **15**, 17, 106
Balazs, Endré, 121
Barraquer, José, 121-22
Bartisch, Georg, 106, 108
Berkeley, George, 63
bifocals, **104**, 105, 106
binocular vision, 69, 103, 108, 137, 145
biomicroscope, 102
bipolar, **29**, 35, 36, **40**, 41, 43, 44, 51, 87, 132
black, 37, 132, 134
black eye, 148
Blakemore, Colin, 60
blind spot, 36-37
blindness, 12, 61, 63, 64, 114, 115, 116, 118, 123
   psychic, see Klüver-Bucy syndrome
   S. B., 64
blue, 32, see also pigment
Boll, Franz, 38
bone structure, 27
Boring, Edwin C., 126-27
braille, 120
brain, 7, 18, **27**, **32**, **45**, 51, 52, 54, 55, 56, **57**, 60, 61, 63, 65, 74, 77, 78, 84, 86, 88, 108, 109, 115, 123
brightness, 17, 52, **53**, 90, 94, **95**, 133
Brunelleschi, Filippo, 137
Brunish, Robert, 28
Bucy, Paul, 54

C
camera, 31, 49, 97, 101, 120, **132**

obscura, 18, **19**
canal of Schlemm, 29, **33**, 114
cataract, 10, 32, 64, 98, 108, **110**, 111, **113**, 114, 118, 121, **122**
catecholamine, 61
cerebellum, **53**, 58, **59**, 60
cerebral cortex, 54, 56
chalazion, 150
chemistry (of vision), 29, 41-43
chiaroscuro, 80
choroid, **29**, 30, 31, 32, **33**, **40**, 41, 107, 116, 121
chromosome, 88
ciliary
   body, 114
   muscle, 32, **33**
   zonule, 32, **33**, 74
Cimabue, Giovanni, 80
colliculus, superior, 54
Collins, Carter, 120
color, 11, 15, 20, **21**, 22, 24, 55, 77-78, 80, 82, 83-84, 85, 86, 87, 88, 90, 92, 94, **95**, 97, see also pigment
   trichromatic theory of, 84, 85, 86
   opponent-process theory of, 84, 86, 87, 97
colorblindness, 87, **88**, 123
   anomalous trichromacy, 88
   monochromacy, 87
   dichromacy, 87-88
computer, 120
cones, 35-36, 37, **40**, 41, 42, 43, 44, 51, 79, 87, 88, 92, 94, 115
conjunctiva, 28, **33**, 114, 149
conjunctivitis, 149
constancy, **70**, 71, see also illusion, size constancy
convergence, 69
Cooper, Grahame, 60
cornea, 15, **24**, 29-30, 31-32, **33**, 64, 102, 103, **104**, 105, 108, **109**, 110-111, 112, 114, **117**, **123**, 148, 151
   donor, 121-122
   transplant, 110-111, **122**
cotton-wool infarcts, 116
cross eyes, see strabismus
cryoprobe, 112, **113**, 116

D
Daviel, Jacques, 111-112
daylight vision, 37, see also cones
de Camp, Michael, **94**
Deering, Michael, 120
de Maupassant, Guy, 98
de Messina, Anotello, 80
Democritus, 11-12, 14
dendrite, **57**
depth, 55, 56, 69
   cues, 65, **66**, 71, 136, 137-38, 139, 141, 142, 145, **146**
   perception, 64, **66**, 72, **72**, 108, 112, 130, 131, 145, **146**
Descartes, René, 63, 74, 83, 90, 108

deuteranopia, 87
dichromacy, see colorblindness
Diocletian, **17**
diopter, 105-06
diplopia, 109
double vision, see diplopia
Dowling, John, 43
drainage canal, 114, 121
drift, 45, 47
drugs, 74
   acetazolamide, 114
   marijuana (THC), 114
   pilocarpine, 114
   Timolol, 114
drusen, 118
duct, 28

E
ear, 49
electrical impulses, **41**, 107
electricity, **29**, 35, 41, 44
electrochemical impulses, **53**, 57
Elements of Psychophysics, 42
elevation of eye, 127
Elizabeth II, **71**
Emerson, Ralph Waldo, 55
Empedocles, 12
endothelium, 111, 112
erythropsin (visual red), 38
Escher, M. C., **124**, 140, **140**, 141
Euclid, 12, 14, 15
evil eye, 9-10, **9**, **10**
eyeball, 7, 51, **102**, 105, 114, 148, 150
eyedrops, 149, 151
eyeglasses, 17, **24**, 32, 105, 106, 118, 122
eye guards, 149
eyelash, 28, 151
eyelids, 28, 108, 148, 149, 150, 151
eye movements, 45-49, 65

F
Fechner, Gustav Theodor, 42-43
Fechner paradox, 43
Fick, A. E., 108
Field, George, **83**
filter, 27, 29, 32, 35, 41, **90**, 144
flick, 45, 47
fluorescein dye, **100**, 103, **109**, 116
   angiogram, 101
focal
   length, 105
   point, **104**
focus, 18, 24, 29, 30, 31, 32, 35, 36, 37, 51, 56, 64, 65, 74, 101, 103, 105, 108, **112**, 122, 133, 137, 145
fovea, **33**, 36, 37, 44, **45**, 47, 48, 49, 56, **104**
Fraser, James, 129
Fraser's spiral, 129-30, **130**
Frey, William, 28
Frisby, John, 52
frontal association cortex, **59**
frontal bone, **33**

Frost, Barrie, 72
Fyodorov, Svyatoslav, 122

G
Galen, 12-13, 14
ganglion cells, **29**, 35, 36, **40**, 43, 44,
    51, 55, 56, 87, 132, 133
Gerstner, Karl, **76**
Gestalt psychology, 61, 129
    principles of perception, 62
glasses, *see* eyeglasses
glaucoma, 151, *see also* pathology,
    glaucoma
glucose, 54
Goethe, Johann, 24, 78
gray, 132, 133
green, *see* pigment
Gregory, Richard, 130
Grosseteste, Robert, 15

H
hallucinations, 74, **74**
Handbook of Physiological Optics, 84
Healon, 121
Hecht, Selig, 38
Held, Richard, 74
Helmholtz-Kohlrausch effect, 97
hemisphere, brain, 53
    left, 52, 54
    right, 52, 54
Hera, **11**
Hering, Ewald, 84, 86, 94
Hermann-Hering grid, 132-33
Herodotus, 10
Hirsch, Helmut, 60-61
Hogarth, William, **139**, 140
horizontal cells, **40**, 43, 51, 87
Horus, **10**
Hubel, David, 51, 55, 56, 58, 61, 62
hue, 78, 80, 82, 83, **84**, 86, 90, 92, 94, **95**
Huygens, Christiaan, 20, 22, **22**, 23
hyperopia, *see* refractive errors,
    farsightedness

I
Iceland spar, **22**
illusion
    Craik-Cornsweet-O'Brien, **133**
    Fata Morgana, 125-26
    geometric, **128**, 133, 143
    McCollough effect, **134**
    moon, 126-29, **126**, **127**
    Müller-Lyer arrows, **135**
    natural, 125-26, 143-45
    Poggendorf, **135**
    Ponzo, **136**
    size constancy, 131, **136**
    space, 145-46, **146**
    subjective contours, **134**
    visual, 126
    waterfall phenomenon, 134
    Zöllner's, **135**
illusionists, 140, **142**

impressionism, 97-98, 123
injuries, 110, 115, 122, 148-49
intraocular pressure, *see* glaucoma
iris, 27, 30, 31, **33**, 38, 42-43, **101**, **102**,
    103, 110, 112, 114, 121
Islamic science, 13-15, **15**

J
Jackson, Hughlings, 74
Jesus, 13

K
kaleidoscope, 129
Kaufman, Lloyd, 127, 129
Kennedy, John, 63
Kepler, Johannes, 14, 18, **19**, 21, 34, 83
Klüver, Heinrich, 54
Klüver-Bucy syndrome, 54
Kohler, Ivo, 74
Kohlrausch, Friedrich, 97
Kühne, Wilhelm, 38, 39

L
lacrimal
    duct, **30**
    excretory duct, **30**
    gland, 28, **30**, **33**
    sac, **30**
lateral geniculate
    body, 56, 58
    nuclei, 52, **53**, **57**, 62
lateral inhibition, 132
Latham, Peter Mayes, 101
Laws, Robert, 71
lens, 15, 17, 18, 20, 24, 30, 31-32, **33**, 34,
    43, 51, 61, 65, 73, 74, 92, 102, 103,
    104, 105, 110, 111-112, **113**, 114, 117,
    121, 126, 132, 143
    convex, **104**, 106, 108
    concave, **104**, 106
    contact, 103, 105, 106, 108, **109**, 118,
        122, 151
    cylindrical, **104**, 106
    intraocular, 112
Leodegarius, St., **15**
Leonardo da Vinci, 9, 13, 14, 17-18, 80,
    108, 139
ligaments, *see* ciliary body
light, 12, 14, 15, 18, 20, 21-25, 31, 32, 36,
    40, 51, **53**, 55, 56, 60, 77, 78, **82**, 83,
    84, 85, 87, 88, **89**, 90, 92, 94, 97, 98,
    101, 102, 103, 105, 106, 110, 111, 122,
    123, 148
    particle theory of, 20, 22, 23
    wave theory of, 20, 22, **22**, 23, 24-25
limbic system, 54
line detectors, 133, 135
linear perspective, 137
Locke, John, 63
Lowell, Robert, **27**
Lucy, St., **16**
luminance, 94
lysozyme, *see* tears

M
MacMillan, Donald B., 125-126
macula lutea, **33**, 36, 116, 118
magic, magician, *see* illusionists
Masaccio, 136, **137**
maxilla, **33**
Maxwell, James, 90
medial canthus, **33**
meiosis, 88
melanin, 30, 31
microscope (-ic), 32, 35, 49, 92
microsurgery, **113**
mirror writing, 74
Mishkin, Mortimer, 54
Monet, Claude, 97-98, **98**
monocular vision, 65, 69, 137, 139
motion, perception of, 43-44, **129**, **131**,
    143
motion parallax, 69
motion pictures, 143
motor cortex, **59**
muscles, 45, **47**, 60, 65, 105, 110, 121
    inferior oblique, **33**, 45
    inferior rectus, **33**, 45
    lateral rectus, **33**, 45
    medial, **33**, 45
    superior oblique, **33**, 45
    superior rectus, **33**, 45
myopia, *see* refractive errors, near-
    sightedness

N
nanometer, 77, **78**, 87
nasal cavity, **30**
nasolacrimal duct, **30**
near point, 32
negative afterimage, 97
Neisser, Ulric, 62
nerve cells, 56, 60
nervous system, **47**
neuron, 55, *see also* nerve cell
neurotransmitters, 51
Newton, Isaac, 20-22, **22**, 23, 82-84, 85
night blindness, 38
night vision, **35**, 37, *see also* rods
    animals, 30

O
occipital lobe, 56, 60
ocular dominance columns, 58, 61
Ocusert, 108
"off center-on surround" cells, 56, **57**
oil-based paint, 80
"on center-off surround" cells, 56, **57**
ophthalmologist, 101, 102, 103, 105, 108,
    113, **116**, **117**, 122, 123, 148, 151
ophthalmology, 101, 120
ophthalmoscope, **24**, **25**, 102-103, 107,
    114, 118, 121
opsin, 41, **41**
optacon, 121
Optica, 12
optic chiasm, 51, 52, **53**, 69

optic disk, **33**, 115
  cupping of, **115**
Opticks, 20
optic nerve, 27, **31**, **33**, 35, 36, **40**, **41**, 44, 51, **53**, **59**, 103, 107, 114, 121, 123, 132
optics, 82, 83, 85, 106, 112, 123
optogram, **38**
optometrists, **24**
orbit, 27, 121, *see also* socket

P
paint, 78-80
Panini, Giovanni Paolo, 139
Papyrus Ebers, 11
Paré, Ambrose, **17**
pathology
  diabetic retinopathy, 116, **117**, **118**
    nonproliferative, 116
    proliferative, 116
  eyestrain, 105
  glaucoma, 102, 103, 105, 111, 114, **115**, **117**, 118
    acute, angle-closure, 114
    chronic, open-angle, 114
  headaches, 105
  retinitis pigmentosa, 115
  retinal detachment, 115-16, **117**, 121
  senile macular degeneration, 116
Peary, Robert E., 125-126
Pecham, John, 17
perception, 62, **70**
peripheral vision, **35**, 112, 114, 115
perspective, 63, **66**, 73, 131, 136-37, 139, 140
phacoemulsifier, 112, **112**
phoropter, 105
photocoagulation, 116
photon, 35, 36, 41, 42
photopic vision, *see* daylight vision
photophores, 143-44
photoreceptor cells, 29, 35, 36, **37**, 38, **40**, 41, 43, 44, 45, 47, 51, **53**, 55, 56, **59**, 84, 86, 87, 132, 133, 143, *see also* rods; cones
pigment, 31, 32, **35**, 36, 38, **40**, 41, 42, **77**, 78, **79**, 80, 86, **89**, 90, 97
pigment epithelium, **40**, 115, 116
pink eye, *see* conjunctivitis
pinhole test, 102
Plato, 12, 14
Pliny the Elder, 38
pneuma theory, 12
polarize (-ation), 23, 92, 143
pons, **59**, 60
Pope, Alexander, 20
posterior chamber, **33**, 114
pressure, *see* pathology, glaucoma
prestriate cortex, *see* visual cortex, prestriate
primary colors, 41, 82, 84, 88, **89**, 98
prism, 20, **21**, 74, 84
protanopia, 87

psychic blindness, *see* Klüver-Bucy syndrome
Ptolemy, 127, 129
Pulfrich pendulum effect, 43-44, **43**
punctum lacrimale, **33**
pupil, 17, 18, 30, 31, **33**, 74, **102**, 103, 114, 151
purple, 35
  visual, *see* rhodopsin

R
radiation in communications, **79**
  infrared, **79**
  X-ray, **79**
  cosmic ray, **79**
rainbow, 30, 84, 88, 90, **92**
rapid eye movement (REM), 48
reading, 62, 63
red, *see* pigment
refraction, 21, 23, 90, 102, 104, **127**
refractive errors, 103, 105, 106, 108, 110
  astigmatism, 103, **104**, 106, 108, 123
  farsightedness, 103, **104**, 105, 106, 121, 122, 123
  nearsightedness, 102, 103, **104**, 105, 106, 121, 122, 123
  presbyopia, 103, **104**, 106
refractive power, 105, 108, 122
Renoir, Auguste, **99**
retina, 7, 18, 21, 25, 29, 30, 31, **33**, 35, 36, 37, 38, **40**, 42, 43, 44, 45, **45**, 47, 48, 51, 52, 53, 54, 56, 58, 62, 65, 71, 73, 84, 86, 87, 94, 97, 102, 103, **104**, 105, 106, 107, 110, 111, 112, 114-15, 118, 121, 126, 130, 131, 132, 133, 134, 143, 145, 146, 148
  torn, 103
retinoscope, 105
rhodopsin, 38, 41-42, **41**
Robert, Etienne Gaspard, 140, 142
Rock, Irvin, 127, 129
rods, 35-36, 37, **40**, **41**, **41**, 42, 43, 44, 51, **79**, 87, 115
Rudolph II, **19**
Rushton, W. A. H., 42, 43

S
saccade, 47-48, **47**, **48**, 49, 146
saturation, 90, 94, **95**
scanning, 48, 62, 64
Scheiner, Christoph, 18
sclera, 29, 30, **33**, **40**, 108, 149
scotopic vision, *see* night vision
sensitivity, 30, 37-38, 42, 133
Seth, **10**
Shiva, **8**
slit lamp, 102, **102**, 103
smooth pursuit, 47, 48
Snell, Willebrord, 21
Snellen, Hermann, 101
Snellen chart, 101, 102
socket, **17**, 27, 28, 45, 150, *see also* orbit
Sokoloff, Louis, 54

sonar spectacles, 120
spectacles, *see* eyeglasses
Spector, Abraham, 111
spectrum, 32, 77, **82**, 84, 85, 87, 88, **89**, 90, 94
Spinelli, D. N., 60-61
spinal cord, **59**
spots (*muscae volitantes*), 35, 116
squint, 109
stereopsis, 52, 108, 110
stereoscope (-scopic vision), 65, **68**, 69, 129
stimulators, 120
strabismus, 109-110, **110**
Stratton, George, 73, 74
striate cortex, *see* visual cortex, striate
stroma, 110
sty, 150
sunglasses, 149
sunlamp, 149
surgery
  couching, 111, **111**
  iridectomy, 112, 114
  keratomileusis, 122
  keratophakia, 121, **122**
  keratotomy (radial), 122-23, **123**
  laser, 114, 116, **117**
  silicone implant, 116
  vitrectomy, 116, 118
Susruta, 111
synapse, 51, **57**

T
Tactile Sensory Replacement (TSR), 120, **120**
tears, 28-29, **28**, 108, 148, 151
telescopes, 20, 23, 49
television, 143
tempera, 80
temporal lobes, 52, **53**, 54
thalamus, **59**
three dimensional vision, 65, 90, *see also* stereopsis
trabecular meshwork, 114
Traber, Zacharias, **18**
Treatise on Light, 23
tremor, 45, 47
trephine, 110
Trevor-Roper, Patrick, 123
trichromacy, anomalous *see* color-blindness
trochlea, **33**
tunnel vision, 114, 118, 146
Tuohy, Kevin M., 108

U
ultrasound (-sonic), **112**, 120-21
ultraviolet light, 32, 35, 49, 103, 109
  burns, 149

V
van Eyck, Hubert, 80
van Eyck, Jan, 80, **81**

vanishing point, **66**, 139
Vasari, Giorgio, 80
vergence, 48-49, *see also* binocular vision
vestibulo-ocular movements, 49
violet, 35
visual acuity, 101-102, 108, 110, 122
visual cortex, 44, 52, 54, 58, **59**, 62, **69**, 132, 133
    striate (primary), 52, **53**, 54, 55, 56, **57**, 58, **59**
    complex cell, 55, **57**, 58
    hypercomplex cell, **57**, 58
    simple cell, 55, **57**, 58
    prestriate (secondary), 52, **53**, 56
visual ray theory, 12, 14, 15, 18, 21, 22
vitamin A, 38, **40**, 41, **41**
vitamin C, **92**

vitreous chamber, 102, 111, 115, 116, **118**
vitreous humor, **33**, 35, **40**, 102, **104**, 112, 121
von Helmholtz, Hermann, **24**, 25, **44**, 84, 97, 107

W
Wald, George, 38, 41-42, **41**
wall eyes, *see* strabismus
Warburg, Otto, 38
wavelength, 31, 41, 77, 78, 84, 85, 87, 88, 90, 92, 94, 144
waves
    light, 77, **78**, 85, **86**, **89**, 94
    radio, 77
Welliver, Jane, **63**
West, Louis Jolyon, 74

Wheatstone, Charles, **68**, 69
white, 37, 132, 133, 134, 136, 144
Wichterle, O., 108
Wiesel, Torsten, 51, 55, 56, 58, 61, 62

Y
Yando, **75**
yellow, 35
Young, Thomas, 20, 24-25, 84, 85, 107, 108
Young-Helmholtz trichromatic theory, *see* color, trichomatic theory of

Z
Zahn, Johann, **22**
Zeus, **11**
Zöllner, Johann, 135

DEC 1 8 1988	DATE DUE	
FEB 2 1983	MAR 9 '88	
May 3	DEC 16	
DEC 13	I	
	NOV 22 '89	
JAN 1 7 1986	FEB 15	
FEB 2 1984		
NOV 22		
APR 23		
FEB 1 0 1986		
MAR 11 1986		